WO... AN... T

JAMES S. HOUSE
University of Michigan

WORK STRESS AND SOCIAL SUPPORT

ADDISON-WESLEY PUBLISHING COMPANY

Reading, Massachusetts
Menlo Park, California • London
Amsterdam • Don Mills, Ontario • Sydney

This book is in the
Addison-Wesley Series on Occupational Stress

Consulting Editor: Dr. Alan A. McLean

Library of Congress Cataloging in Publication Data

House, James S 1944–
Work stress and social support.

(Addison-Wesley series on occupational stress; 4)
Bibliography: p.
Includes index.
1. Job stress. 2. Job stress—Social aspects.
3. Stress (Psychology) 4. Helping behavior. I. Title. II. Series.
HF5548.85.H67 613.6′2 80-22234
ISBN 0-201-03101-9

ISBN 0-201-03101-9

ABCDEFGHIJ-AL-89876543210

To my parents, for their early
and enduring social support

FOREWORD

The vast literature concerned with the individual coping with work stress stems from many and diverse disciplines, primarily psychiatry, clinical and social psychology, sociology, cultural anthropology, and occupational and internal medicine, with significant contributions from such widely different fields as behavioral toxicology and personnel and management. While each discipline is concerned with so-called "psychosocial stressors," communication between the several disciplines has generally been the exception rather than the rule. Lawyers, for example, tend to communicate mainly with other lawyers about the issues that concern them. Union leaders tend to communicate most often with other union leaders. Clinical psychologists direct their communications to their colleagues, but use a different language from that used by many of the psychiatrists who are equally concerned. Even social psychologists and industrial sociologists sometimes find it difficult to exchange data. The transfer of useful data from one discipline to another has proven to be very difficult. "Some researchers go about rediscovering the known, with little deference to an existing literature or to determinable frontiers for contemporary research; and what consensus may be possible is *not adequately disseminated for beneficial application beyond home base.*"*

---

* Robert Rose, editorial, *Journal of Human Stress,* Vol. 3 No. 1, March 1977.

Communication across disciplines is not the only difficulty that students of job-related stress encounter. Transcultural communication is a problem too. Western physiologists, for instance, who are concerned with hormones in the brain, have difficulty communicating with their eastern European colleagues who prefer to speak in terms of "higher nervous function."

There is growing common concern. Theories and practices in each discipline are beginning to cross-pollinate other disciplines and to exert a positive influence toward understanding the stresses of the workplace and workers' reactions.

The many denominators of concern for an employee population under stress form the unifying theme of these volumes. As a field of study, occupational stress is beginning to gel. It is a subject of increasing interest not only to members of unions and management, but also to the health professionals who serve as their consultants. Increasingly, awareness and expertise are being focused on both theoretical and practical problem solving. The findings of social scientists have led to the enactment of legislation in the Scandinavian countries, for instance, where employers are now required, under certain circumstances, to provide meaningful work and appropriate job satisfaction with a minimum of occupational stress.

The authors of these books represent many points of view and a variety of disciplines. Each, however, is interested in the same basic thing—greater job satisfaction and greater productivity for each employee. The books were written independently with only broad guidelines and coordination by the editor. Each is a unique, professional statement summarizing an area closely related to the main theme. Each extracts from that area applications which seem logically based on currently available knowledge.

All of the authors treat, from differing perspectives, three key concepts: stress, stressor, and stress reactions. *Stress* defines a process or a system which includes not only the stressful event and the reaction to it, but all the intervening steps between. The *stressor* is a stressful event or stressful condition that produces a psychological or physical reaction in the individual that is usually unpleasant and sometimes produces symptoms of emotional or physiological disability. The *stress reaction* concerns the consequences of the stimulus provided by a stressor. It is, in other words, the response to a stressor, and it is generally unhealthy. Most often, such reactions may be defined in rather traditional psychological terms, ranging from mild situational anxiety and depression to serious emotional disability.

Many frames of reference are represented in this series. A psychoanalyst describes the phenomenon of occupational stress in executives. A sociologist reflects the concern with blue-collar workers. Health-care-delivery systems and the prevention of occupational stress reactions are covered by occupational physicians. Other authors focus on social support systems and on physiological aspects of stress reactions. All the authors are equally concerned with the reduction of unhealthy environmental social stimuli both in the world of work and in the other aspects of life that the world of work affects. In each instance, the authors are concerned with defining issues and with drawing the kinds of conclusions that will suggest constructive solutions.

The legal system, beginning with worker's compensation statutes and more recently augmented by the Occupational Safety and Health Act, deals directly with occupational stress reactions and will be the subject of one of the books in the series. That statute, which created both the Occupational Safety and Health Administration and the National Institute for Occupational Safety and Health, contains a specific directive mandating study of psychologically stressful factors in the work environment. We have seen criteria documents and standards for physical factors in the work environment. We may soon see standards developed to govern acceptable levels of psychological stressors at work such as already exist in Sweden and Norway; another significant area of concern for this series.

At the beginning of this series it is difficult to foresee all the pivotal areas of interest which should be covered. It is even more difficult to predict the authors who will be able and willing to confront the issues as they emerge in the next few years. In a rapidly changing technological, scientific, and legislative world, the challenge will be to bring contemporary knowledge about occupational stress to an audience of intelligent managers who can translate thoughts into constructive action.

Alan A. McLean, M.D.
Editor

ACKNOWLEDGMENTS

Writing is work, and as such is at times stressful as well as fulfilling. It is especially fitting in a book about work stress and social support to acknowledge the sources of emotional, instrumental, informational and appraisal support that have made writing this book less stressful and more fulfilling than might otherwise have been the case. The initial idea for this volume, and the series of which it is a part, was Alan McLean's. While contributing to my work stress in pushing for an early completion of the manuscript, he also provided prompt editorial feedback and support for my efforts.

I have profited greatly from discussions and collaborative research on social support with colleagues and students, first at Duke University and the University of North Carolina, and later at the University of Michigan, which has proved an especially stimulating environment for writing such a volume. John R. P. French, Jr., Theodore Newcomb, and Ronald Kessler not only discussed issues of methods and substance, they also provided detailed and constructive critiques of the manuscript. Arthur Shostak and David Bertelli provided similarly valuable reviews on behalf of Addison-Wesley. I have been fortunate to work also with two insightful and industrious students in the course of my research on social support—James A. Wells at Duke and James M. LaRocco at Michigan. Both have contributed to the development of my thinking, and Jim LaRocco also critiqued a draft of this book. Berton N. Kaplan played an important role in initiating some of this research as well as commenting on the manuscript. My task was much aided by Sidney

Cobb's and the late John Cassel's early syntheses of social support literature, and Sid also kindly reviewed a next-to-final draft. Toni Antonucci, Robert Caplan, and Robert Griffin also provided input at different points along the way; Bertil Gardell offered a Swedish perspective; and Halsey House, a manager's view.

While writing this book, I and my family have also been settling into a new place. Three people have especially eased both of these processes for me. Robert L. Kahn made a detailed commentary on the next-to-final draft, and more generally has modeled both how to give social support and how to write about it. Marie Klatt typed the initial outline for the volume as one of her first secretarial activities with me. Since then she has not only typed and retyped the manuscript with great care, skill, and good humor, but also taken major responsibility for numerous other aspects of the manuscript preparation—from drawing figures to obtaining permissions to generating indices. Her efforts on this book and other tasks have made them all much easier and more enjoyable for me. My wife, Wendy, has been a source of constructive suggestions for this book, especially on issues of application, and more importantly, a true kindred spirit with whom to share the ups and downs of a busy and full stage of life.

Finally, a number of people whom I have never seen facilitated the successful completion of this project. Charles T. Peers, Jr., provided overall editorial direction, while Gail Gangi and Sue Zorn took care of numerous publishing details. The editorial efforts of Elydia Siegel contributed significantly to both the readability, and timely completion, of this volume.

The permission of the following publishers to reprint several tables and figures is hereby gratefully acknowledged: The American Sociological Association, *Canadian Journal of Behavioral Science*, and the *American Journal of Epidemiology*.

Ann Arbor, Michigan J.S.H.
October 1980

INTRODUCTION

The subject of this book—the role of social support in reducing work stress and improving health—constitutes an important frontier both scientifically and practically. The study of social support in relation to stress and health is a rapidly developing, but not yet well-developed, area of theory and research at the interface between the social and biomedical sciences. Thus, one could write a volume addressed solely to scientists on the existing state and future directions of theory and research on social support. But social support also has important implications for practical efforts to reduce stress, protect health, and enhance the quality of working life. Thus, one could also write a volume addressed solely to managers and practitioners on how to apply existing knowledge about social support to the operation of work and work organizations in western society. This volume seeks, however, to speak to both of these audiences and sets of concerns.

This dual focus stems from more than indecision on my part. Any separation between basic research and theory on the one hand and applied research and practice on the other is artificial, especially in an area in which neither basic research and theory nor applied research and practice are well developed. Kurt Lewin once said that there is nothing as practical as a good theory, meaning that efforts to change and improve aspects of social life make sense, and are likely to succeed, *only if* they are derived from a logical theory, the implications of which have been tested and supported in sound empirical research. Both individually and collectively we have been victimized too often in the past two decades by people promising to change our lives for the better using programs that lacked a sound scientific foundation. The result has often been an expensive personal or social investment with little positive, and sometimes a negative, outcome. Thus, those concerned primarily with application and practice must also be concerned and informed about the scientific foundation on which proposals for application and practice rest. The more scientific portions of this book seek not only to enlighten scientists but also to give informed laymen a foundation for evaluating the potential utility of proposed applications of this knowledge of work stress, social support, and health.

Conversely, I would argue that practical application constitutes a crucial test of the validity of scientific knowledge and an important stimulus for the further development of theory and research. Many of the most important areas of social science, including social support, cannot be definitively studied, or even simulated, in the laboratory. The real world must be the laboratory in which social scientific knowledge is finally tested. The ultimate test of scientific understanding of social support, for example, will come from field experiments in which researchers attempt to apply their understanding in real life settings *and* to evaluate the results of these efforts using the most rigorous scientific criteria that can be ethically and practically employed. The results of such field experiments will have as much, and probably more, impact on the development of further theory and research as any program of more basic research. The single most important stimulus to the development of basic theory and research in modern social science was probably the first intensive effort to apply social science during World War II. Thus, it behooves those who are concerned primarily with the scientific issues in this volume to think about the practical implications of these scientific issues *and* about the scientific implications of the practical and applied issues discussed here.*

By its dual focus this volume seeks to promote closer contact and cooperation of more basic scientists with applied researchers

---

* I do not mean to argue that all theory and research should have immediate practical application, only (1) that research of a more applied nature can and often does yield "basic" insights and (2) that the ultimate validation of "basic" knowledge comes from its practical application.

and practitioners. This volume brings together a body of scientific evidence that argues for applied attempts, initially on an experimental basis, to enhance social support as a means of reducing occupational stress and/or protecting and improving the physical and mental health of workers. I hope this argument is sufficiently convincing to those with decision-making power over working conditions that they will want to undertake such experiments and to insure that they both have a sound base in prior scientific knowledge and will receive adequate scientific evaluation. Similarly, I hope that scientists interested in the problem of social support will be convinced of the need to direct some of their effort toward planning and evaluating such field experiments. Mutual interchange between the science and practice of social support should produce both better scientific understanding and a higher quality of working life.

A SOCIAL STRUCTURAL FOCUS

I approach the study of work stress, social support, and health from the perspective of a sociologist (or really a sociological social psychologist). The unique focus of sociology as a discipline is on the collective nature of social life, or what is sometimes termed social structure. The study of work stress, social support, and health is inherently a social psychological problem, involving an interplay between the nature of individuals and the nature of the social environments and social structure in which they are enmeshed. Hence, the genesis of stress and the alleviation of it and its effects can involve either the nature of individuals and/or the nature of their social structures or environment.

A clinical approach to the problem of stress, however, tends to focus primarily on the individual, since it is with individual cases that clinicians must ultimately deal. Such a focus is also quite compatible with the individualistic ethos of our American society. We need to recognize, however, that the source of, and solution to, many problems of stress and health can be, and sometimes must be, social or structural rather than individual. Although clinicians ultimately deal with individual cases, scientists, managers, and/or policy makers are generally more concerned with collective trends and outcomes. Similarly, they must be concerned with collective causes of such trends or outcomes. This volume will emphasize the ways in which the social environment and social structure contribute to, and can alleviate, problems of work stress and health.

Sociologists use the term social structure to refer to recurring patterns of behavior by some set of individuals. Thus, social structures come in many sizes and levels depending on the number of individuals involved. All people are enmeshed in patterns of face-to-face interpersonal relationships at home and at work, and these structures continually affect their lives. In Chapter 2 the definition of social support involves describing one or more aspects of these patterns of interpersonal relationships. Individuals and more intimate social structures are also importantly affected by the larger social structures of which they are a part—the organizations (and subunits thereof) in which they work, their communities, and even the national economy. The concern here will be with how these various structures affect the levels of stress, social support, and health experienced by the individual.

In sum, the emphases of this book necessarily reflect the orientation of its author as a research sociologist. I know that more practically and clinically oriented persons may find my concerns with issues of theory and research at times too academic for their taste. However, I also know that my efforts to speak to a lay audience about practical problems will leave my scientific and academic colleagues feeling at times that I am making assertions without sufficient scientific justification or qualification. This dilemma is embodied in the comments I received from two readers on sections of the initial draft of this book:

> *As a scientist you're OK; as a salesman, a flop!* (from a clinically oriented physician)
>
> *Sometimes you seem overeager to demonstrate that social support really* can *(so glad you say "can" rather than "does") buffer effects of stress on health. In so far as other readers get this impression it could lead to questions about your scientific objectivity* (from an academic social psychologist).

I feel that having to struggle to balance issues of science and practice has ultimately been rewarding for me; this volume attempts to say some worthwhile things in both domains.

CONTENTS

I

STRESS, SOCIAL SUPPORT, AND HEALTH: THEORETICAL FOUNDATIONS

1

WORK STRESS, SOCIAL SUPPORT, AND HEALTH: THE PROMISE AND THE PROBLEMS

A major goal of life and work is to achieve, maintain, and enhance physical and mental health and the quality of life itself. People's concern for health is manifest in the prestige, power, and financial resources they accord to the medical profession and health care institutions in the United States. This concern is also evident in the burgeoning popular interest in learning how to live more happily and healthfully. For centuries physical, chemical, and biological agents in the environment were considered the primary threats to health, and improvements in health and well-being stemmed from a combination of therapeutic and preventive measures for combatting these agents and their effects. From the midnineteenth to the midtwentieth centuries biomedical scientists and practitioners revolutionized their ability to understand and control infectious disease, enhancing both the length and quality of people's lives.

In the past few decades, however, the primary threats to health have changed, and accomplishments in combatting them have been more modest. The major sources of serious physical illness and death are now the so-called chronic diseases, especially heart disease and cancer, which tend to develop and progress over a relatively long period of time in response to a wide range of causal factors, rather than being precipitated suddenly by a specific agent or germ. Social and psychological factors have been increasingly recognized as among the important factors in the etiology of a wide range of physical disorders. Furthermore, in this century the concept of health has steadily broadened to include mental as well as physical well-being. Yet to a great extent health professionals have tried to deal with chronic diseases and mental illness in much the same way they dealt with infectious diseases, through a combination of preventive and therapeutic procedures directed at their physical, chemical, and biological causes and consequences.

The result has been growing discontent with both the costs and the benefits of the health care system and especially with the cost–benefit ratio. There is no question that progress has been made, but it has largely been progress in developing more effective therapies for treating disease rather than in the prevention or postponement of the onset of disease. Further, the increased control of disease produced by new therapies often comes at a substantial cost in both money and side effects of treatment that detract from the quality of life. Thus, it is clear that we must seek ways to prevent the onset of disease, which will necessarily involve increased attention to the social, psychological, and behavioral factors in the etiology of both mental and physical health problems.

STRESS, HEALTH, AND SOCIAL SUPPORT

The term social stress encompasses many of the social, psychological, and behavioral factors in health and disease. Available evidence indicates that the stress process may contribute to the development of a wide range of physical and mental disorders, including infectious diseases, chronic respiratory ailments, cardiovascular diseases, gastrointestinal disorders, depression, and perhaps even cancer (cf. Levine and Scotch, 1970; Syme, 1974; Cassel, 1976; Lipowski, Lipsitt and Whybrow, 1977, Totman, 1979). Thus, a major preventive strategy for improving physical and mental well-being is to intervene in the stress process that contributes to these disorders. Such a strategy has several parts that may be pursued separately or together. A more detailed discussion of the stress process that links potential stressors to short- and long-term stress reactions will be presented in Chapter 2.[1] For the moment, three

1. To maintain consistency with the terminology of the rest of this series on occupational stress, the term *stressor* is used here to refer to the actual or objective nature of a situation such as a job. I term the perception of such situations *perceived stress.* The term *stress* is used rather loosely to refer to stressors or perceived stresses or to the larger stress process outlined in Fig. 2.3.

major elements of a strategy for preventing deleterious effects of stress on health may be noted. First, the level of stresses which have deleterious health consequences can be reduced. Second, inputs that bolster health (or other outcomes that may be adversely affected by stress) can be provided, and hence compensate for deleterious impacts of stresses. Third, the impact of stresses on health can be mitigated or buffered.

For example, if work-related stressors make many individuals depressed, several things could be done. First, the work-related stressors could be reduced or eliminated through changes in the work situation (including encouraging people who feel extremely stressed at work to change jobs). Second, efforts could be made to enhance the nonwork aspects of the people's lives. These efforts would tend to decrease the average level or rate of depression to some extent, even if the work situation remained unchanged, thus compensating in part for the effects of work-related stress. The general health status of people is a function of a mix of both deleterious and healthful forces, and any increase in deleterious forces may be compensated for by an increase in beneficial forces. Finally, the impact of work-related stressors on health could be mitigated or buffered in some way. That is, the effect of stressors on health is blocked or reduced by doing something that necessarily neither reduces the level of stressors nor improves health, but prevents stressors from having a deleterious effect on health. This "something" then would allow people to be exposed to stressors, with less likelihood that their health would be adversely affected.

For example, excessive exposure to the sun has adverse effects on people's skin. These adverse effects can be reduced by (1) staying out of the sun, or (2) treatments and practices that enhance the health of the skin (e.g., good habits of nutrition and cleanliness, perhaps the application of certain moisturizing agents, taking of vitamins), or (3) use of sunscreens which block the effect of damaging solar rays on the skin. The crucial difference between (1) and (2) or (3) is that the latter two can have beneficial effects on health without reducing exposure to the stressor, while the first one improves health only to the extent that exposure to the stressor is reduced. The difference between compensating and buffering factors is that the compensating factors have beneficial effects on health regardless of exposure to stressors whereas buffering factors, by definition, have beneficial effects on health only among people exposed to stressors. Thus the application of sunscreens will not improve the health of the skin of people who have little exposure to the sun, whereas the compensating strategies may improve the dermatological well-being of people regardless of exposure to the sun.

Clearly, any or all of these strategies are desirable and promising routes to improving health. As discussed in Chapter 6, levels of exposure to health hazards, including stress, should be reduced wherever feasible. Similarly, practices that promote health and well-being should be encouraged. However, efforts to directly reduce stress or improve health entail costs as well as benefits, and appreciable levels of stress may be inherent in many life and work situations. Mechanisms for buffering the impact of stress on health are thus especially significant because they offer the only feasible means of protecting and promoting health in situations where (further) stress reduction or health promotion are impossible or prohibitively expensive.

Where that point is in any given situation is difficult to pinpoint, but such limits clearly exist. To take another analogy closer to our concern with work stress, the National Institute of Occupational Safety and Health (NIOSH) and the Occupational Safety and Health Administration (OSHA) are charged by the Occupational Safety and Health Act of 1970 with responsibility for establishing and ensuring "safe" levels of exposure to physical, chemical, and biological hazards in the work environment. Both the determination of what constitutes "safe" levels of exposure, and the development of means of reducing exposure to these hazards has, however, proved a complex and difficult process. Similarly, there are limits to what people can do to live healthfully. Thus, the only viable strategy in many instances is to develop and make available mechanisms that can ameliorate or eliminate (i.e., buffer) the deleterious impact of these hazards on health. Most often this buffering is done, as in the case of exposure to the sun, by blocking or lessening the actual impact of these hazards on workers through the use of protective devices—radiation suits, respirators, goggles, safety helmets, ear muffs. The provision of such protective devices is relatively straightforward, if somewhat expensive, although their effective use is more problematic.

Scientists neither know nor are likely to find any simple or effective means of preventing irreducible social stressors from impinging on people in the way that they can prevent irreducible radiation, or noise, or dust from actually contacting workers bodies. Air traffic controllers are generally considered to experience a good deal of stress in their work, and they manifest heightened levels of a number of psychosomatic diseases such as hypertension and ulcers. Although the current levels of stress among air traffic controllers could certainly be reduced, the experience of substantial stress, at least at times, will probably always characterize this and many other occupations. Thus something must be found that can lessen or buffer the impact of occupational stress on health.

Social and biomedical scientists have recently become intrigued with something termed *social support* that may have precisely these properties. Social support also appears capable of reducing the level of at least some occupational stressors and of directly promoting aspects of health as well. It does not promise to be a panacea for all problems of stress and health. But the quantity and quality of people's social relationships with spouses, friends, coworkers, and supervisors appear to have an important bearing on the amount of stress they experience, their overall well-being, and on the likelihood that stress will adversely affect their overall well-being. These social relationships have been neglected in some ways in thinking about planned interventions to improve the quality of life and health. Yet, especially in the work environment, it appears possible, at least in theory, to enhance social support without great expenditures of time and money and with few negative side effects.

Thus social support provides an attractive and, so to speak, triple-threat strategy for preventing deleterious effects and promoting beneficial effects of the social environment on health. The interest in social support is evident both in scholarly writing on the topic (e.g., Cassel, 1976; Cobb, 1976; Caplan and Killilea, 1976) and in efforts to apply these ideas in health policy. Most notably community support systems were a central focus of the report and recommendations of the President's Commission on Mental Health (1978a–d). The Commission (1978a, p. 15) identified "personal and community supports" as "one of the most significant frontiers in mental health" and proposed as their very first recommendation: "A major effort be developed in the area of personal and community supports which will: (a) recognize and strengthen the natural networks to which people belong and to which they depend; . . ." The same conclusions could apply to physical as well as mental health.

WORK STRESS AND SOCIAL SUPPORT: THE PROMISE

For a variety of reasons, work and work organizations can and should constitute a major focus of efforts to improve health by strengthening social supports. As McLean (1979, p. 1) and others (e.g., House, 1974b) have emphasized, the time and importance most adults invest in their work suggests that what happens on the job can have pervasive effects on their health and well-being. The effect of work on health was recognized as an area of national policy in the Occupational Safety and Health Act of 1970. The focus of this legislation was on physical, chemical, and biological hazards in the work environment, but it also mentions the potential importance of psychosocial factors. And a growing body of evidence indicates that psychosocial forms of occupational stress have deleterious effects on a wide range of physical and mental health outcomes, including a possible tendency to intensify the noxious effects of some hazardous physical-chemical agents (Caplan, Cobb, French, Harrison, and Pinneau, 1975; House, 1974a, 1974b; House and Jackman, 1979; House, McMichael, Wells, Kaplan, and Landerman, 1979; Kasl, 1974, 1978; McLean, 1979). Thus, occupational stress and its effects on health must constitute a major focus of any effort to utilize social support to reduce stress or improve health.

On other grounds as well, work and work organizations constitute an attractive focus for any preventive approach to health, and particularly one focusing on social support. The time spent at work and the organizations in which that time is spent constitute the most structured and organized aspects of most adults' lives. (The same could be said of the role of schools in children's lives). Thus, work and work organizations are potentially powerful and efficient mechanisms of planned social intervention and change. That is, for any given input of time, money, and manpower researchers can potentially produce a greater impact on more normal or premorbid adults through the workplace than through any other single formal organization or institution—a point that is insufficiently recognized by many people concerned with health promotion or disease prevention programs, including the President's Commission on Mental Health.

Finally, social support, especially from others at work, is an especially appealing mechanism for dealing with the problem of occupational stress and health. Efforts should be made to reduce levels of occupational stress wherever such efforts are feasible and effective, which is more often than commonly thought. Nevertheless, people cannot or may not want to reduce certain types and instances of occupational stress, because the associated costs of doing so are prohibitive or because some types and degrees of stress may have non-noxious or even beneficial consequences. Stress is ubiquitous and not always a bad thing, as Hans Selye (1974) has tried to indicate with his notion of *eustress* (stress that is beneficial and even pleasant rather than harmful and unpleasant). Further, stress reduction is only one of many goals that an organization or an individual worker may have. A special appeal of social support is that it may buffer people against the impact of the irreducible crises and stresses of work. As will be discussed in Chapter 6, enhancing social support at work may also have a number of ancillary benefits for workers and work organizations besides improvements in worker health.

Occupational stress is a significant health problem, and attempts to enhance social support at work constitute a potentially effective yet efficient mechanism for reducing work stress, improving health, and buffering people against the effects of work stress on health. The work organization provides a good mechanism for efforts to enhance support, and enhanced support may improve individual and organizational effectiveness beyond its effects on stress or health or the relationship between them.

WORK STRESS AND SOCIAL SUPPORT: THE PROBLEMS

A number of qualifying words—appears, probably, potentially—have been used repeatedly in this introductory discussion, and with good reason. Although social support is a promising idea that can potentially be applied in a wide range of ways to improve the quality of life and work, it is also in many respects a new idea that has been conceptualized differently by different people. It has not yet been explored extensively in empirical research nor applied widely in programs of stress or disease prevention. Thus, this book cannot be a straightforward presentation of well-established theoretical propositions, empirical facts, or methods of putting these propositions and facts to practical use in promoting health and preventing disease. Rather, it must be more an exploration and assessment of new and varied ideas, scattered empirical facts, and selected (sometimes untested) practical applications of available knowledge. I hope that this exploration will produce a useful, if only approximate, map of some virgin territory, indicating to both scientists and persons involved in real work organizations: (1) what is known and what needs to be known about social support in relation to work stress and health, and (2) what are the most reasonable implications of existing theory and knowledge for both private and public policies regarding the nature of work and work organizations.

GOALS OF THE BOOK

Specifically this book has four major goals, each of which is the focus of one or two of the remaining chapters:

- Chapter 2 seeks to clarify the meaning of the vague term *social support* and to indicate how and why support should or could reduce stress, improve health, or buffer the impact of stressors on health. Since "nothing is as practical as a good theory," researchers need to develop a theory of support to guide both scientific research and practical applications.
- Chapters 3 and 4 review and evaluate available evidence (negative as well as positive) regarding the idea or hypothesis that social support does reduce stress, improve health, and buffer the impact of stress on health. No matter how well-stated and logical the theory may be, people would hesitate to act on it unless its implications had been borne out in empirical research.
- Chapter 5 reviews and evaluates what is known about the determinants or causes of social support. This topic has been neglected in recent writings on social support that have concentrated on the effects of support rather than on the factors that promote or inhibit its development. Yet such knowledge is crucial if practical efforts to reduce stress or improve health by enhancing social support are to be made.
- Finally, Chapter 6 assesses the practical implications of the theoretical ideas and empirical data discussed in Chapters 2, 3, 4, and 5. Do researchers know enough to develop practical applica-

tions or interventions? If so, what do existing theory and data suggest as to the nature of such applications and interventions? And how can and should we evaluate (do research on) extant or planned applications or interventions so that both future theories and practice will be improved on the basis of previous experience?

2

THE NATURE OF SOCIAL SUPPORT

Social support, like stress, is a concept that everyone understands in a general sense but it gives rise to many conflicting definitions and ideas when we get down to specifics. We all have an intuitive sense of what constitutes social support: we know people who we regard as "supporters" or "supportive" of us, from whom we receive support, and to whom we often give it. Parents are generally our earliest sources of support, augmented and eventually supplanted by friends, relatives, spouses, children, and various people with whom we have more limited and specific relationships—work supervisors and associates, ministers, teachers, physicians, nurses, and counselors. What we get from, and sometimes give to, these people are roughly those things that a dictionary defines as the act of support: ". . . to keep from falling, slipping or sinking . . . give courage . . . or confidence to; help; comfort . . . give approval to . . . maintain or provide . . . with money or subsistence . . . vindicate or corroborate . . . to keep up; maintain; sustain . . ." (Webster's New World Dictionary, 1959, p. 1465). Huckleberry Finn, for example, received social support from Tom Sawyer, the widow Douglas, and his black companion, Jim, among others, but the kinds of support received from each of them were quite different, as were the consequences.[1]

1. Alan McLean (personal communication) has rightly suggested that people may achieve a sense of support from familiar activities, places, and things, such as fishing or rafting on the Mississippi for Huck Finn. The focus here will remain on social support or support from people since it is more prevalent, variable, and manipulable in most individual's lives.

Although social support is a relatively new topic in discussions of stress and health in general, and work stress and health in particular, what is denoted or connoted by the term *social support* is hardly new (Cobb, 1976, p. 301). Social support has been implicitly or explicitly central in earlier literary, religious, sociological, psychological, and medical thought; it has just had different names: love, caring, friendship, a sense of community, and social integration. Thus, in some ways social support is really old wine in a new bottle. What is distinctive about this wine bottled under the social support label, however, is the claim that support may reduce stress, improve health, and, especially, buffer the impact of stress on health.

For some purposes, even initial scientific exploration, a general and intuitive understanding of the notion of social support may be sufficient. To develop a level of scientific understanding adequate to the formulation of effective, practical applications, however, we must achieve greater precision and insight in our understanding of social support and its relation to work stress and health. Both the current literature on social support that is reviewed next and the dictionary definitions already noted suggest many different types or components of social support. But what are the differences that make a difference? That is, how many types or forms of social support must be distinguished that have distinctive and important effects on work stress, health, or the relation between them? Although love may make the world go 'round, it would not be of interest here if it did not affect work stress or health or the relation between them.

Researchers also must be able to determine the extent to which a person is receiving (or giving) various types of social support. That is, we must be able to measure social support if we are to study its effects and causes and ultimately learn how to enhance social support so as to reduce stress, improve health, or buffer the relation between stress and health. Finally, we must understand more about the mechanisms through which social support operates. As noted in Chapter 1, if social support improves people's health, it could do so by reducing their exposure to stress, by improving their ability to adapt to stress, or improving health directly. And it could do each of these things in several ways.

Thus, if we are to see and capitalize on the potential scientific and practical uses of the concept of social support we must first know what it is, how to measure it, and how it operates. This chapter seeks to provide some initial answers to these three questions—answers that seem reasonable and logical. Succeeding chapters will evaluate these answers in light of existing data and evidence.

WHAT IS SOCIAL SUPPORT?

Some expert opinions

One way to find out how to define and measure social support is to ask scientists who have studied social support. Surprisingly, the experts are sometimes vague or circular, even contradictory. Cassel (1976), for example, in one of the two major reviews of the impact of social support on stress and health, provides no explicit definition of social support.[2] Lin, Simeone, Ensel, and Kuo (1979, p. 109), essentially define social support as support that is social: "Social support may be defined as support accessible to an individual through social ties to other individuals, groups, and the larger community." Such imprecision in conceptions of support is mirrored in operational measures that are conglomerations of anything that might protect people against stress and disease, including ego strength and social class (Nuckolls, Cassel, and Kaplan, 1972), or job satisfaction (Lin et al., 1979). Studies with such measures can be highly misleading, since the results may reflect primarily effects of ego strength, social class, or job satisfaction rather than social support.

Others have generated more explicit and appropriate, if somewhat disparate, definitions of social support. The translation of these definitions into operational measures for research, evaluation, and practice is just beginning. Sidney Cobb (1976, p. 300) begins his major review paper on social support by defining *social* support as:

> *. . . information belonging to one or more of the following three classes:*

2. A paper by Kaplan, Cassel, and Gore (1977) reviews many different conceptions and definitions of support, but neither identifies one of them as better than the others nor proposes a new definition of support. The present discussion parallels aspects of the structure and content of the Kaplan et al. paper.

1 *Information leading the subject to believe that he is cared for and loved;*
2 *Information leading the subject to believe that he is esteemed and valued;*
3 *Information leading the subject to believe that he belongs to a network of communication and mutual obligation.*

Here and in a later paper Cobb (1979) refers to these three aspects of social support as: (1) "emotional support," (2) "esteem support," and (3) "network support." In his later paper, Cobb (1979, pp. 93–94) explicitly distinguishes *social* support from: (a) "instrumental" support or counseling, (b) "active" support or mothering, and (c) "material" support or goods and services. Although Cobb is correctly distinguishing among different types of support and focusing attention on the most important, labelling only one type as social support is unduly restrictive.

In contrast, Kahn and Antonucci (1980) define social support as "interpersonal transactions that include one or more of the following key elements: *affect, affirmation,* and *aid.*" They go on to define affect as "expressions of liking, admiration, respect, or love," including under one heading what Cobb terms "emotional" support and "esteem" support. Affirmation refers to "expressions of agreement or acknowledgment of the appropriateness or rightness of some act or statement of another person." Finally, aid refers to "transactions in which direct aid or assistance is given, including things, money, information, time and entitlements," which Cobb chooses to label as "material," "active," and/or "instrumental" support rather than as social support.

These two discussions suggest considerable consensus about the general nature of social support, but considerable disagreement over specifics. Other authors only add to the variety of definitions of support and its component elements somewhat further. Pinneau (1975, p. 2) distinguishes among tangible, appraisal (or information), and emotional support:

> Tangible *support is assistance through an intervention in the person's objective environment or circumstances, for example: providing a loan of money or other resources. . . .* Appraisal *or* information *support is a psychological form of help which contributes to the individual's body of knowledge or cognitive*

system, for example: informing the person about a new job opportunity, explaining a method for solving a problem. . . . Emotional *support is the communication of information which directly meets basic social-emotional needs, for example: a statement of esteem for the person, attentive listening to the person. . . . The term* psychological support *may be used to subsume both appraisal and emotional support.*

Gerald Caplan has written widely on the role of "support systems" in relation to stress and especially community mental health (Caplan, 1974; Caplan and Killilea, 1976). He defines support systems as:

. . . attachments among individuals or between individuals and groups that serve to improve adaptive competence in dealing with short-term crises and life transitions as well as long-term challenges, stresses, and privations through (a) promoting emotional mastery, (b) offering guidance regarding the field of relevant forces involved in expectable problems and methods of dealing with them, and (c) providing feedback about an individual's behavior that validates his conception of his own identity and fosters improved performance based on adequate self-evaluation.

(CAPLAN AND KILLILEA, 1976, p. 41)

Caplan sees support systems as including "professions and formal community institutions as well as natural systems (such as the family; nonprofessionalized and informal social units, particularly mutual aid organizations; and person-to-person caregiving efforts both spontaneous and organized)" (Caplan and Killilea, 1976, p. 41).[3]

Perhaps because these authors feel that they have a good intuitive sense of the meaning of social support, they have developed largely deductive or a priori definitions of support. The result is some consensus about the range of aspects of relationships that are within the general domain of social support, but considerable disagreement as to which of these aspects are most important.

---

3. To define support in terms of its effects, as Caplan does, begs the major question for research and practice—how, when, and for whom are supportive social relationships beneficial in adapting to stress. A better definition might read: "attachments . . . that *may* serve to improve adaptive competence. . . ."

Views of ordinary people

A complementary approach would be to ask ordinary people to describe actual relationships in which they are involved and which they consider supportive. Gottlieb (1978) has done just this with a sample of 40 single mothers receiving social assistance in Canada. These women were asked in semistructured interviews to identify several "personal and important problems which you are experiencing and which are of concern to you," and the persons who had been helpful to them in dealing with these problems. For each such person respondents were asked: (a) "How has X become involved in helping you deal with the problem or your feelings about it?", and (b) "Is there anything about X as a person or about his/her way of dealing with the problem that stands out for you?" Table 2.1 presents a classification of the responses developed by four trained coders of the interview protocols.

The classification in Table 2.1 serves two main purposes. First, it constitutes a record from everyday experience against which the a priori theoretical schemes just discussed can be evaluated. Second, it provides specific and concrete illustrations of what is meant by social support and the aspects of it discussed above. That is, the examples in Table 2.1 indicate what we do or say to others and what they do or say to us that is perceived as helpful or supportive in dealing with the problems of living we all face. Gottlieb's first major category of "emotionally sustaining behavior" appears to subsume what Cobb terms "social support," what Kahn and Antonucci term "affect" and "affirmation," what Pinneau terms "emotional support," and what Caplan describes as "promoting emotional mastery" and perhaps "feedback . . . that validates . . . (the person's) identity and fosters improved performance based on adequate self-esteem." Gottlieb's second major category of "problem-solving behaviors" subsumes aspects of Cobb's "instrumental," "active," and "material" supports, Kahn and Antonucci's "aid," Pinneau's "appraisal" and "tangible" supports, and Caplan's "guidance." Gottlieb's small third category, termed indirect personal influence, appears to be merely a generalized, nonproblem-specific form of "emotionally sustaining behavior"; and, Gottlieb's final category, "environmental action," merely an environmentally oriented form of "problem-solving behaviors." In sum, Gottlieb's classification fits reasonably well with

TABLE 2.1 *A classification scheme of informal helping behaviour*

Category		Definition	Example
A. EMOTIONALLY SUSTAINING BEHAVIOURS			
A1	Talking (unfocused)	Airing or ventilation of general concerns without reference to problem specifics	"she'll talk things over with me"
A2	Provides reassurance	Expresses confidence in R as a person, in some aspect of R's *past* or *present* behaviour, or with regard to the future course of events	"he seems to have faith in me"
A3	Provides encouragement	Stimulates or motivates R to engage in some *future* behaviour	"she pushed me a lot of times when I was saying, 'oh, to heck with it' "
A4	Listens	Listening only, without reference to dialogue	"he listens to me when I talk to him about things"
A5	Reflects understanding	Signals understanding of the facts of R's problem or of R's feelings	"she would know what I was saying"
A6	Reflects respect	Expresses respect or esteem for R	"Some people look down on you; well, she doesn't"
A7	Reflects concern	Expresses concern about the importance or severity of the problem's impact on R or for the problem itself	"just by telling me how worried or afraid she is" (for me)
A8	Reflects trust	Reflects assurance of the confidentiality of shared information	"she's someone I trust and I knew that it was confidential"
A9	Reflects intimacy	Provides or reflects interpersonal intimacy	"he's just close to me"
A10	Provides companionship	Offers simple companionship or access to new companions	"I've always got her and I really don't feel alone"
A11	Provides accompaniment in stressful situation	Accompanies R in a stressful situation	"she took the time to be there with me so I didn't have to face it alone"
A12	Provides extended period of care	Maintains a supportive relationship to R over what R considers an extended period of time	"she was with me the whole way"

TABLE 2.1 *A classification scheme of informal helping behaviour* (Cont.)

Category	Definition	Example
B. PROBLEM-SOLVING BEHAVIOURS		
B1 Talking (focused)	Airing or ventilation of specific problem details	"I'm able to tell him what's bugging me and we discuss it"
B2 Provides clarification	Discussion of problem details which aims to promote new understanding or new perspective	"making me more aware of what I was actually saying other than just having the words come out"
B3 Provides suggestions	Provides suggestions or advice about the means of problem solving	"he offered suggestions of what I could do"
B4 Provides directive	Commands, orders, or directs R about the means of problem solving	"all Rose told me was to be more assertive"
B5 Provides information about source of stress	Definition same as category name	"she keeps me in touch with what my child's doing"
B6 Provides referral	Refers R to alternative helping resource(s)	"financially, he put me on to a car mechanic who gave me a tune-up for less than I would pay in a garage"
B7 Monitors directive	Attempts to ensure that R complies with problem-solving directive	"making sure that I follow through with their orders"
B8 Buffers S from source of stress	Engages in behaviour which prevents contact between R and stressor	"he doesn't offer it (alcohol) to me anymore"
B9 Models/provides testimony of own experience	Models behaviours or provides oral testimony related to the helper's own experience in a similar situation	"just even watching her and how confident she seems has taught me something"
B10 Provides material aid and/or direct service	Lends or gives tangibles (e.g. food, clothing, money) or provides service (e.g. babysitting, transportation) to R	"he brought his truck and moved me so I wouldn't have to rent a truck"

B11	Distracts from problem focus	Temporarily diverts R's attention through initiating activity (verbal or action-oriented) unrelated to the problem	"or he'll say, 'Let's go for a drive' . . . some little thing to get my mind off it"
C. INDIRECT PERSONAL INFLUENCE			
C1	Reflects unconditional access	Helper conveys an unconditional availability to R (without reference to problem-solving actions)	"she's there when I need her"
C2	Reflects readiness to act	Helper conveys to R readiness to engage in future problem-solving behaviour	"he'll do all he can do"
D. ENVIRONMENTAL ACTION			
D1	Intervenes in the environment to reduce source of stress	Intervenes in the environment to remove or diminish the source(s) of stress	"she helped by talking to the owners and convincing them to wait for the money a while"

Note: R = Respondent (the people interviewed by Gottlieb)
Source: Gottlieb (1978)

prior definitions and conceptions of support, while also illustrating and clarifying the more abstract ideas in those conceptions.

Toward an adequate conception of support

At this point researchers can identify the major issues or questions that an adequate conception of social support must address, and can provide some tentative answers to these questions. These issues can be expressed in the question: *Who* gives *what* to *whom* regarding *which* problems? The full range of potential forms of social support can be defined in a matrix such as Table 2.2. The columns of Table 2.2 indicate the potential persons or groups *who* may provide support, while the rows indicate *what* kinds of support these persons may provide. (The *whom* and *which* of support are discussed in the section on measurement of social support.)

The entries in the columns include the major individuals or groups that may provide social support, ranging from informal sources (family and friends) to persons connected with major life roles such as work to professional or semiprofessional persons and groups who provide specific services, including forms of support. These include what are termed, following Gottlieb, service or care givers. Persons who provide major services to an individual or a family (child care and domestic workers, legal or financial advisors) can be significant sources of social support. People may also seek out individuals or groups for specific kinds of support, including the self-help or mutual help groups such as Alcoholics Anonymous or Parents without Partners, and professionals in health (mental or physical) and welfare services (physicians, nurses, social workers, psychologists, clergy). Some analyses of social support (Caplan and Killilea, 1976; President's Commission, 1978) have emphasized more formal sources of support (professionals, self-help groups), while others (Cassel, 1976; Cobb, 1976) have emphasized more informal sources (family, friends, coworkers). A thorough assessment of support will consider all possible sources, although some sources will be more important than others depending on the nature of the person and problem needing support.

Informal and nonprofessional sources of support (family, friends, coworkers, especially work-related supports) will be the focus of this book for several reasons. First, these sources are much more commonly mentioned when people are asked to name

TABLE 2.2 *Potential forms of social support*

	Source of Support								
Content of supportive acts	(1) Spouse or partner	(2) Other relative(s)	(3) Friend(s)	(4) Neighbor(s)	(5) Work supervisor	(6) Coworker(s)	(7) Service or care giver(s)	(8) "Self-help" group(s)	(9) Health/Welfare professional(s)
1. *Emotional Support* (esteem, affect, trust, concern, listening)									
2. *Appraisal Support* (affirmation, feedback, social comparison)									
3. *Informational Support* (advice, suggestion, directives, information)									
4. *Instrumental Support* (aid in kind, money, labor, time, modifying environment)									

Within this matrix of types of social support, each can be:

(a) *general* versus *problem-focused*
(b) *objective* versus *subjective*

actual sources of support, and this tendency has increased in recent years (cf. Gottlieb, 1978; Gurin, Veroff, and Feld, 1960; Veroff, Douvan, and Kulka, in press). Second, these sources are the most truly preventive forms of social support in that, if effective, they preclude the need for more formal support or treatment. Effective prevention programs must enhance the effectiveness of these informal supports, although the need for and utility of more and better professional services remains. Finally, the data reviewed in succeeding chapters suggest these informal sources can be very effective in reducing stress, improving health, and buffering the impact of stress on health.[4]

The rows of Table 2.2 define four broad classes or types of *supportive behaviors or acts.* These four classes derive from the conceptions already reviewed, but none of the these conceptions includes all four types. Rather than trying to decide which of these is really social support (as Cobb does), all should be considered as potential forms of support and their impact on stress and health (and the relation between these) treated as an empirical question. Just as the relevance of sources of support varies with the person and problem requiring support, so would the relevance of different types of support. Nevertheless, *emotional support,* which involves providing empathy, caring, love, and trust, seems to be the most important. Emotional support is included in one form or another in all the schemes reviewed, and its impact on stress and health is clearly documented in succeeding chapters. When individuals think of people being "supportive" toward them, they think mainly of emotional support; this category subsumes the largest number of the specific acts of support reported by Gottlieb's (1978) respondents.

The other forms of support in Table 2.2 are less universally recognized in theoretical discussions, and less conspicuous in reports of ordinary people. Nevertheless, each may be important in specific instances.[5] *Instrumental support* is the most clearly dis-

---

4. Informal sources may be uniquely effective for two reasons. First, informal relationships are usually based on mutual respect and relative status equality. Support may be more meaningful and more easily accepted when coming from a respected peer. Second, informal support is usually a spontaneous act, rather than a role-required behavior (of a counselor, clergyperson, etc.). Thus, such support may seem more genuine and hence be more effective.

5. Whereas it is likely that the effects of emotional support are almost always positive, tending to reduce stress and improve health, the other forms

tinguished from emotional support, at least in theory, involving instrumental behaviors that directly help the person in need. Individuals give instrumental support when they help other people do their work, take care of them, or help them pay their bills. It is important to recognize, however, that a purely instrumental act also has psychological consequences. Thus, giving a person money can be a sign of caring or a source of information and appraisal (very possibly of a negative variety, communicating to persons that they are in need of money and dependent on others for it).

Appraisal and informational support are the most difficult to clearly define and distinguish from other forms of support. *Informational support,* means providing a person with information that the person can use in coping with personal and environmental problems. In contrast to instrumental support, such information is not in and of itself helpful, rather it helps people to help themselves. For example, informing an unemployed person of job opportunities or more generally teaching them how to find a job is informational support. Obviously, providing information may imply emotional support and may, at times, constitute instrumental support (for example, if the person's major need is information) as in the case of tutoring or coaching vocational or academic knowledge.

Appraisal support, like informational support, involves only transmissions of information, rather than the affect involved in emotional support or the aid involved in instrumental support. However, the information involved in appraisal support is relevant to self-evaluation—what social psychologists have termed social comparison (Festinger, 1954; Jones and Gerard, 1967, Ch. 9). That is, other people are sources of information that individuals use in evaluating themselves. Such information can be implicitly or explicitly evaluative. Work supervisors, for example, may tell workers that they are doing good (or poor) work, or they may tell

---

of support may sometimes increase stress or impair health. Thus Cobb (1979) notes that instrumental support can make people overly dependent on others, and John R. P. French Jr. and James LaRocco (personal communication) have noted that informational and appraisal support may increase people's perceptions of stress (in ways that may or may not be realistic). Workers, for example, can validate and even accentuate each other's feelings of work stress and dissatisfaction. It is also useful to recognize that support may have different effects in the short run versus the long run. Honest feedback or appraisal is sometimes painful, but beneficial in the long run.

workers what constitutes the performance of an average worker and let them decide for themselves whether they are above or below average.

Each of the four types of support can be classified into more specific acts or types of support (see Table 2.1), but these four constitute a minimal set of potential types of social support that is more inclusive than any of the conceptions previously addressed. This typology will undoubtedly be improved upon in the future, but it should be useful in thinking about both research and practical application. And thinking about such problems should improve future conceptual understanding.

MEASURING SOCIAL SUPPORT

Social support, then, is a flow of emotional concern, instrumental aid, information, and/or appraisal (information relevant to self-evaluation) between people. The goal here is to understand how such flows affect stress, health, and the relation between stress and health. To understand such flows, researchers must be able to measure support—to determine how much of what kind of support person *P* is receiving or person *O* is giving. Any attempt to measure support, however, raises further definitional issues.

Subjective or perceived support versus objective support

To determine how much emotional support you get from your spouse or work supervisor, I might ask you: "How much is (your spouse or your work supervisor) willing to listen to your work-related problems?" Or more generally: "To what extent is your (spouse or work supervisor) concerned about your welfare?" [6] Alternatively, I might ask your spouse or supervisor how much each of them is "willing to listen to your work-related problems"

6. Such questions have been asked and are presented in Table 4.1. These questions focus on work and are only two of many possible questions regarding emotional support. I might also ask about instrumental support (e.g., helping you do your work better), informational support (providing you with information you need to do things you want to do), or appraisal support (providing you with honest feedback about yourself or your work). Researchers are able to measure and document the effects of emotional support. The ability to measure and show important effects of the other forms of support has yet to be clearly established.

or is "concerned about your welfare." Finally I might observe your interaction with your supervisor or spouse and decide for myself how much each of them is willing to listen to your work-related problems or is concerned about your welfare. No study of social support, stress, and health has measured support in these three different ways and compared the results, although students of marital and family relations have often looked at how family members' perceptions of their relationships compare with each other and with those of observers from outside the family. Students of social support could profit from using a similar strategy on occasions.

Studies of social support have most often asked people to rate how much emotional support they are receiving from others—asking you, for example, how much your supervisor or spouse is concerned about your welfare. The resulting answers are usually termed *subjective* or *perceived* support. This method is at once the easiest and most appropriate procedure in initial efforts to understand the relation of social support to stress and health. It is easiest because researchers can collect data on social support from the same person from whom they are collecting data on stress and health. It is also appropriate because social support is likely to be effective only to the extent it is perceived. That is, no matter how much your spouse or supervisor feels or acts supportive toward you, there will be little effect on you unless you, in fact, perceive them as supportive.

Increasingly, however, researchers will need to get what are often termed measures of objective support—ideally the observations of one or more trained scientific observers using methods of proven reliability and validity. A simpler substitute for such scientific observations is to ask how much support your supervisor or spouse or a neutral observer feels you are receiving.[7] To enhance the amount of support you feel you receive from your supervisor or spouse, requires understanding of how what these people think,

---

7. These measures are objective only in the sense of being relatively independent of your subjective perceptions of support. You and your spouse or supervisor will often have different perceptions of the same events, and it is often difficult or impossible to say one person is clearly more objective (in the sense of fair, accurate, or unbiased). Neutral third parties are more likely to be objective in this latter sense as well, but still lack training in scientific methods of observation.

feel, or do affects your perception of support. Thus the relation between objective and subjective perceived support is an important but neglected topic taken up in more detail in Chapter 5. For now, as indicated in Table 2.2, simply remember that measures of support may be subjective or objective (see Caplan, 1979).

Problem-focused versus general support

A second issue that arises in attempts to measure support is whether to treat support as *general* or *problem focused*—for example, asking whether others are "concerned about your welfare, in general," or are "willing to listen to your *work-related* problems" (with no necessary implications about willingness to listen to your problems more generally). There can be no hard or fast rules here. But if the crux of understanding social support is to understand *who* gives *what* to *whom* about *which problems,* it is clear that the who and what depend on the whom and which. That is, certain sources and types of support will be more relevant to some types of people (whom) and problems (which). Gottlieb (1978) for example, has shown that single mothers predominantly use informal sources of support (family, friends, neighbors) in dealing with most of their problems, but are especially likely to use more formal sources (care givers) in dealing with child-oriented problems. Similarly, he shows that these women feel emotional support is most relevant to dealing with personal–emotional problems, while what I have termed instrumental support is most relevant for dealing with both child-oriented and financial problems.

The focus of this book is on occupational stress. Hence, supervisors and coworkers are more important sources of support here than they were for Gottlieb's single mothers, and some types of social support will be more consequential than others, depending on the particular occupational stressors of concern. Thus, Table 2.2 notes that in any situation the measures of support may be more general or more problem focused. To the extent that they are problem focused, only some sources or types of support may be relevant.

Measures of support: ideal versus actual

Table 2.2, then, provides a general framework for the definition and measurement of social support. To attempt to measure all aspects of support suggested in Table 2.2 would be impossible and fruitless in any single situation or study. What the table does is to summarize the issues that a priori seem likely to be important to understanding social support. Discovering which are important is the task of research and practice. As concerns move to such concrete research and practice, however, thinking tends to be more pragmatic and less conceptually abstract and differentiated. Research and practice evolve from initially crude measures and manipulations of support to increasingly refined ones.

A necessary condition for supportive acts or behaviors is some interaction between two people. Further, since providing social support to another person involves some costs in time, energy, and sometimes money or goods, most people neither give nor receive social support totally gratuitously. Giving or receiving social support usually involves expectations of reciprocity. Thus flows of social support occur primarily in the context of relatively stable social relationships rather than fleeting interactions among strangers. The people we intuitively think of as sources of support are not people we barely know, but our parents, spouses, friends, work associates, or professionals with whom we have continuing relationships.[8]

A minimal condition for experiencing social support, then, is to have one or more stable relationships with others. Hence, the crudest approach to the measurement of support is to measure the quantity and structure of social relationships. Many studies of social support use marital status, for example, as their sole or primary measure of support. Others also determine whether relationships exist with friends, neighbors, relatives, or more formal organizations. Some go further and measure the frequency of interaction or contact as well as the mere existence of the relationships (see Chapters 3 and 4). Finally, rather elaborate analyses are sometimes made of people's "social networks," yielding measures of the size, density, connectedness, and homogeneity of people's social relationships (see Kaplan, Cassel, and Gore, 1977; Mitchell, 1969).

No matter how sophisticated they may be, however, such methods usually consider only whether a relationship exists or not,

8. The degree to which flows of social support are symmetrical or reciprocal is an interesting issue addressed in later chapters.

and say nothing about the nature, quality, or content of a relationship. The information obtained in such studies is, for example, whether a person is married or not, rather than whether or not the marriage is happy and meets various needs of the person. It is unlikely, perhaps impossible, that an unhappy marriage provides as much or more support as a happy one, yet precisely this assumption is made whenever all that is assessed is whether the person is married or not, or how many friends or contacts they have, or what organizations they belong to. On logical and intuitive grounds a more adequate conception and measure of support will consider not only the structure of a person's social relationships, but also the content or quality of those relationships. This may be achieved by asking people to identify for us only those relationships that they perceive as supportive or by asking them whether each of their nominal "friends," coworkers, or family members is, in fact, supportive.

The measurement of social support must move, and has moved, increasingly toward assessment of the dimensions or aspects of relationships such as those discussed in connection with Table 2.2. Finally, if people are to develop more supportive relationships, they will need to understand how specific acts or behaviors contribute to the development of such relationships. Researchers are encouraged to pursue these more difficult tasks of measurement and analysis, however, because even crude measures of support have yielded substantial evidence that social support can reduce stress, improve health, and buffer the relation between them.

MECHANISMS OF SOCIAL SUPPORT

Main effects versus buffering

Chapter 1 suggested that social support is a potential triple threat in dealing with problems of work stress and health. That chapter briefly indicated the three major ways in which social support can reduce stress and improve health and drew some analogies between these potential effects of support and efforts to reduce the effects of physical-chemical hazards in the environment. Let us now consider more explicitly how and why social support appears to be such a promising means of dealing with problems of stress and health.

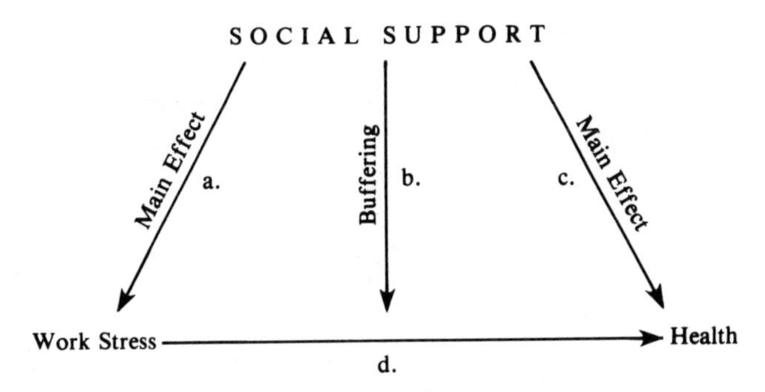

FIG. 2.1 *Potential effects of social support on work stress and health*

Fig. 2.1 graphically illustrates the three ways in which social support can affect work stress and health. The figure and this book assume, based on a growing body of evidence (Kasl, 1978; McLean, 1979; House and Jackman, 1979) that occupational stress has deleterious effects on mental and physical health (arrow *d* in Fig. 2.1). Support can modify or counteract this deleterious effect in three ways. First, social support can directly enhance health and well-being because it meets important human needs for security, social contact, and approval, belonging, and affection (arrow *c* of Fig. 2.1). That is, positive effects of support on health can offset or counterbalance negative effects of stress.

Second, support, at least from people at work, can directly reduce levels of occupational stress in a variety of ways (arrow *a*), and hence indirectly improve health (via arrow *d*). For example, supportive supervisors and coworkers can minimize interpersonal pressures or tensions; and the experience of support can satisfy work-related motivations for affiliation, approval, and accurate appraisal of the self and environment, generally leaving workers more satisfied with themselves and their jobs.[9]

These two effects of social support, which might be called

9. As already noted, the effects of emotional support are most likely to have a uniformly positive effect on health and a uniformly negative effect on stress levels. Certain forms of instrumental, informational, and appraisal support may have adverse effects in some circumstances, although the preponderance of effects would undoubtedly be health enhancing and stress reducing.

main effects, are important, but also obvious to many people. Recent interest in social support has been sparked mainly by a third type of effect—the potential of social support to mitigate or buffer the impact of occupational stress (and other social stresses) on health (arrow *b* in Fig. 2.1). Here social support has no direct effect on either stress or health, but rather modifies the relation between them, much as a chemical catalyst modifies the effect that one chemical has on another.

The concept of buffering is implicitly or explicitly central in most of the major writings on social support and some authors have gone so far as to suggest that buffering is virtually the only way in which support affects health. Thus, Cobb (1976, p. 300) entitled his 1976 Presidential Address to the Psychosomatic Society "Social Support as a Moderator of Life Stress" and focused on "the way that social support acts to prevent the unfortunate consequences of crisis and change." He goes on to assert that "one should not expect dramatic main effects (on health) from social support" (Cobb, 1976, p. 302). Similarly, Caplan's definition of social support systems cited above emphasizes their role in improving "adaptive competence in dealing with short-term crises and life transitions as well as long-term challenges, stresses, and privations . . ." (Caplan and Killilea, 1976, p. 41). The implication of these statements is that the deleterious impact of stress on health is mitigated (or even eliminated) as social support increases, and conversely that support will have its strongest beneficial effect on health among people under stress and may have little or no beneficial effect for people not under stress. Kaplan et al., (1977, p. 49) makes this implication explicit: "social supports are likely to be protective only *in the presence* of stressful circumstances."

Despite the scientific prominence and potential practical importance of the idea of buffering, considerable confusion exists about what constitutes evidence of buffering versus main effects of social support, whether existing empirical data provide such evidence, and whether it matters if support has buffering effects or main effects. It is important first to understand the meaning of main versus buffering effects of social support and why this distinction is important, before considering what conclusions are supported by existing empirical evidence.

The need to distinguish main versus buffering effects arises when considering how stress and social support may combine to affect health. Fig. 2.2 illustrates three possible ways in which social support and stress may jointly affect health. Each graph in Fig. 2.2 depicts the linear relationship between occupational stress and health for three different levels of social support: low(···), medium(---) and high(——).

Fig. 2.2(a) illustrates a pure buffering effect. Note that occupational stress has no effect on health if people have high social support, but the effects of stress on health become increasingly adverse as support decreases. Viewed another way, social support has no beneficial effect on health among persons with little stress, but the beneficial effects of support become increasingly apparent as stress increases. In contrast Fig. 2.2(b) illustrates a case where

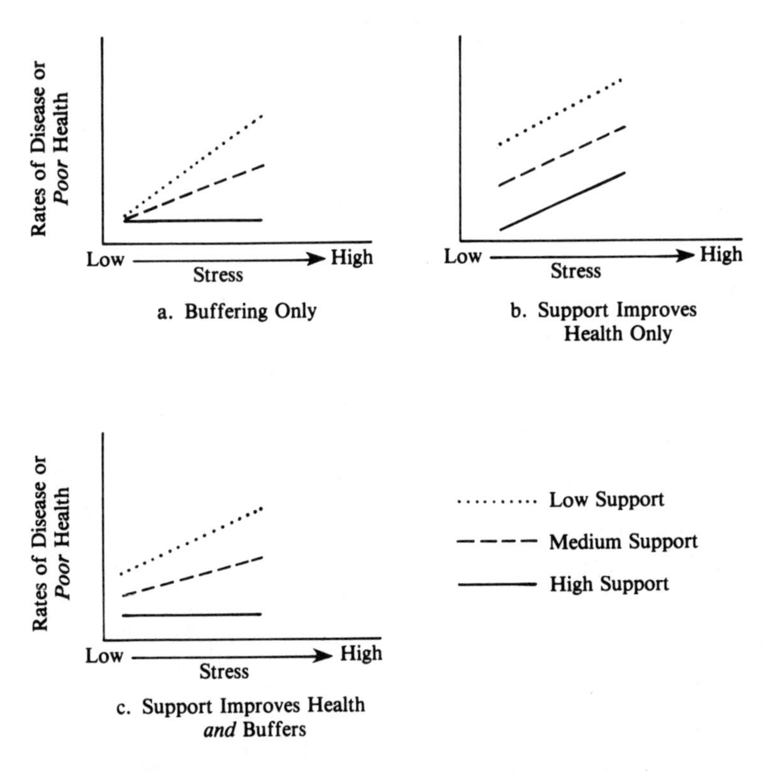

FIG. 2.2 *Different patterns of effects of social support and occupational stress on health*

support has a main effect on health, but the slope of the relationship between stress and health is unaffected by levels of support, indicating no buffering. Regardless of stress level, rates of disease or poor health increase as social support decreases. Finally, Fig. 2.2(c) illustrates the case where support has both a main effect on health and a buffering effect on the relationship of stress to health. Note that the three lines are not parallel, indicating a buffering effect, but that even when stress is low people with high support are healthier, indicating a main effect of support on health.[10]

Although the basic distinctions depicted in Figs. 2.1 and 2.2 are straightforward, there has been considerable confusion about how to make these distinctions in empirical research. This confusion has in turn led to theoretical confusion. For example, in a turnabout from his 1976 paper, Cobb (1979, p. 99) suggested it is "not worth worrying about the distinction between main effects and interaction effects." I would argue that it is important to recognize this distinction and to test for such differences in empirical research, while recognizing that the theoretical and practical importance of social support does not rest solely on its having one or another kind of effect. Much writing on social support has implicitly or explicitly equated support effects with buffering, and this view is clearly too narrow. Nevertheless, the buffering effects of support are one of its most intriguing and important properties, and important practical implications follow from knowing whether the effects of support are primarily main effects or primarily buffering effects or a mixture of both.

To the extent that support has largely main effects, that is, it acts to reduce stress (regardless of levels of health) and to improve health (regardless of levels of stress), everyone would benefit from enhanced levels of social support. To the extent, however, that support has primarily buffering effects, it will be of significant value to people experiencing moderate to high levels of stress, but of lesser, or even no, value to people experiencing little or no stress. Assuming that in many cases the resources available for

---

10. What is shown graphically in Figs. 2.1 and 2.2 is translated into mathematical-statistical form in Appendix A. The one point from Appendix A worth noting here is that buffering effects are statistical interaction effects between stress and support in predicting health, while the main effects are simple additive relationships.

enhancing social support are limited, to the extent that the effects of support are primarily buffering effects, efforts to enhance support should be directed primarily at high-stress groups, much as the distribution of a scarce vaccine is targeted on high-risk groups in the population.

In fact, support has both kinds of effects. Its buffering effects are especially unique, however, and offer a strategy for alleviating the unhealthy consequences of irreducible stresses at work. Although much can and should be done to reduce stress at work, stress is as certain a component of human life as death and taxes. If people are to function well as individuals, families, organizations, and as a society they must learn to live with and adapt to stress. To the extent that it has buffering effects, social support can help us in this regard. McLean (1979) and others have pointed to the way of life of oriental societies as a source of ideas for how to adapt to stress. The focus has been on oriental philosophy of life and oriental modes of individual response to the world (e.g., meditation). But, as indicated in succeeding chapters, Americans may also have something to learn from the social structure of oriental society about how to facilitate the giving and receiving of social support.

Understanding the difference between buffering versus main effects is only the beginning of understanding the mechanisms through which social support comes to affect stress and health. If we are to utilize social support effectively in adapting to a world of stress, we must understand in more detail how it comes to affect stress and health in the case of both main and buffering effects.

Social support and the stress process

The term *stress* has been used in so many confusing and contradictory ways that Cassel (1976, p. 108) concludes that "the simple-minded evocation of the word stress has done as much to retard research in this area as did the concepts of the miasmas at the time of the discovery of microorganisms." Nevertheless, an increasing number of researchers (McGrath, 1970; Levine and Scotch, 1970; French, Rodgers, and Cobb, 1974; Kagan and Levi, 1974) have converged on a similar conceptualization of the nature of stress as a phenomenon or process. Fig. 2.3 presents a paradigm of stress research (from House, 1974a) that reflects this convergence (and is consistent with the definition of stress as a process in McLean, 1979).

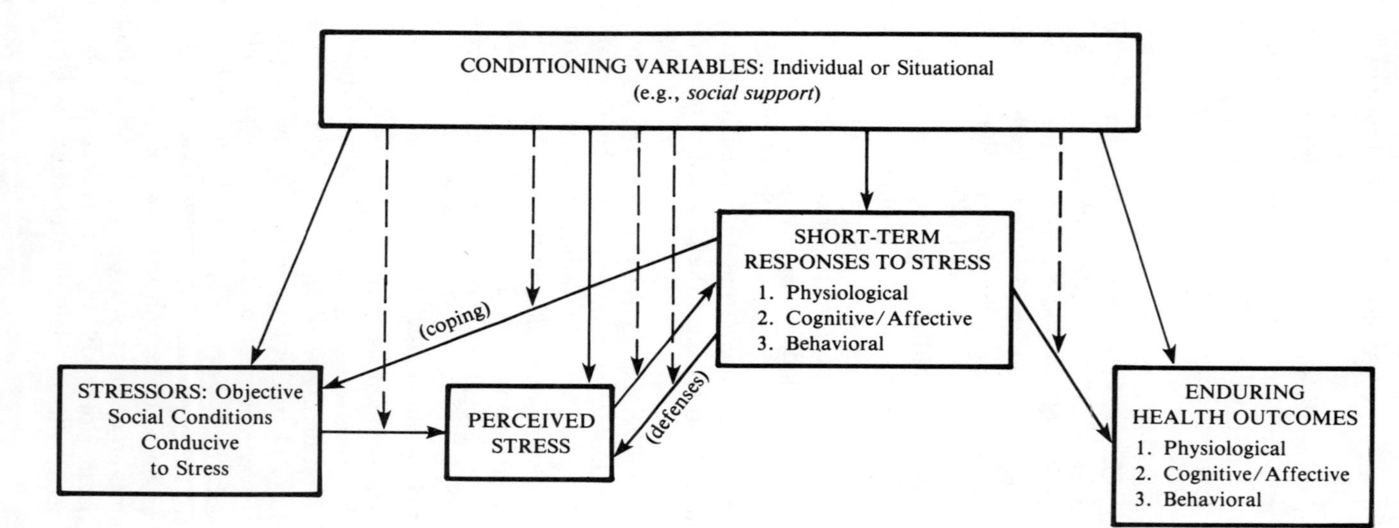

NOTE: Solid arrows between boxes indicate presumed causal relationships among variables. Dotted arrows from the box labeled "conditioning variable" intersect solid arrows, indicating an interaction between the conditioning variables in the box at the beginning of the solid arrow in predicting variables in the box at the head of the solid arrow.

FIG. 2.3 *A paradigm of stress research (Adapted from House, 1974a).*

The solid arrows in this paradigm represent potential main (additive) effects, while the dotted arrows indicate potential buffering (or interactive) effects. Social support, for reasons indicated above, may have main effects that reduce the level of objective stressors or perceived stress and improve health. For example, a supportive environment contains fewer objective conditions conducive to stress such as interpersonal conflict. Being loved, cared for, or simply listened to can make people feel better, thus reducing perceived stress and also promoting (directly or indirectly through its effects on stress) physical and mental health. Support may similarly affect the level of short-term responses or more enduring outcomes such as anxiety, heightened blood pressure, or increased alcoholic drinking.[11] Less obvious, perhaps, are the potential ways in which support may buffer the impact of occupational stress on health.

The paradigm of Fig. 2.3 posits that people perceive stress in response to certain objective social conditions. Conditions are usually perceived as stressful when the demands on people exceed their abilities or when they are unable to fulfill strong needs or values (McGrath, 1970; French et al., 1974). That is, people's needs or abilities are incongruent with their social environment in significant and consequential ways. Except perhaps for extreme situations such as disasters or concentration camps, however, no objective social or occupational situation will *necessarily* produce perceptions of stress or resultant physiological, psychological, or behavioral responses and outcomes in all people exposed to the situation. Rather, how people perceive a given situation depends on other individual or situational factors, labeled conditioning variables in Fig. 2.3, of which social support is one. Considerable evidence from social psychological experiments indicate that the presence of other people alters initial perceptions of objective social stimuli (Lazarus, 1966; Tajfel, 1969). Thus, social support could mitigate or buffer the effect of potentially stressful objective situations (such as a boring job, heavy workloads, unemployment)

---

11. This book focuses on the role of support in helping to prevent occupational stress and physical and mental disorders, but it is worth noting that a considerable literature shows that support can facilitate the treatment of disease, for example, by increasing compliance with medical regimens (see Cobb, 1979, p. 98; Caplan, Robinson, French, Caldwell, and Shinn, 1976).

by causing people initially to perceive the situation as less threatening or stressful and hence leading them to manifest less of those psychological, physiological, or behavioral responses productive of disease (see Cassel, 1976; Kagan and Levi, 1974, for discussions of how psychosocial stress produces physiological, as well as psychological and behavioral, responses and outcomes).

Even if a situation is initially perceived as stressful, however, social support may still lessen or eliminate the tendency of this perceived stress to lead to responses productive of disease. Fig. 2.3 indicates that once a situation is perceived as stressful a variety of responses are possible, some of which may serve to modify the objective social conditions (arrow labeled coping) and/or the person's perception of it (arrow labeled defense) so as to reduce or eliminate the perception of stress and hence to alleviate its impact on health. Cobb (1976, p. 311) has suggested that social support helps to buffer persons against stress primarily by facilitating efforts at coping and defense. Table 2.1 provides numerous examples of how social support helps people to adapt to the stresses of daily life.

Where efforts at coping and defense fail to reduce the objective stressors or the perception of stress, social support may still alleviate the impact of such perceptions on short-term physiological, psychological, or behavioral responses that can produce more enduring health or disease outcomes. Without altering the perception of stress per se, social support may reduce the importance of this perception to individuals and hence their degree of reaction to it. Research shows, for example, that support from spouses may mitigate the impact of job dissatisfaction on health by helping the person to recognize that the job is not so important in the total context of life and that dissatisfactions with it may be compensated for by satisfactions and accomplishments outside of work. Support may also have some kind of general tranquilizing effect on the neuroendocrine system, making people less reactive to perceived stress. Findings of social support effects in studies of animals suggest such a process (Bovard, 1959; Cassel, 1976). Finally, supportive others may facilitate healthful behaviors (for example, exercise, personal hygiene, proper nutrition, and rest) that increase individuals' abilities to tolerate or resist psychosocial stress and physical, chemical, or biological threats to health, as well.

SUMMARY AND CONCLUSION

This chapter has set the stage for later chapters by discussing what social support is, how it can and should be measured, and how and why it can and should affect work stress and health. Both scientific experts and relatively uneducated laypersons agree that social support is an interpersonal transaction involving one or more of the following: (1) emotional concern (liking, love, empathy), (2) instrumental aid (goods or services), (3) information (about the environment), or (4) appraisal (information relevant to self-evaluation). Thus, there are at least four different types of support —emotional, instrumental, informational, and appraisal—that are important to distinguish because they have different causes and consequences. It is the task of research and practice to determine the relative importance of these different types of support in affecting stress and health. However, both experts and laypersons tend to believe emotional support is the most important, a view confirmed by empirical research in succeeding chapters.

Both to conduct research and to use social support as a tool for reducing stress and improving health, we must be able to measure it. Thorough measurement will indicate *who* gets *how much* of *what kinds* of support *from whom* regarding *which* problems. Initial efforts to measure support, however, have been rather crude—often asking merely whether people have social relationships with others. For example, are they married, do they have friends, do they belong to organizations? Even such crude measures suggest that support does reduce stress and improve health. More refined measures will allow researchers to identify the specific amounts and kinds of support and the specific sources of support that are most helpful in alleviating specific kinds of problems such as occupational stress and health. Better measurement is also important for practical reasons. Effective applied intervention must seek to influence the content and quality of important social relationships (for example, marriage or work relationships) rather than just establishing or disestablishing relationships. We often have little control over whether people have such relationships, but we can try to make the relationships they do have more supportive.

Social support can affect stress and health through a variety of mechanisms. Most interesting here is the distinction between main effects of support on stress and health and buffering effects of support on the relationship between stress and health. That is, supportive social relations may reduce the experience of stress and improve health. More importantly, support appears to buffer or mitigate the impact of stress on health. Thus it can be an important factor in helping people to live more healthfully in the presence of stress, which is an inevitable and sometimes irreducible element of work and life. Chapters 3 and 4 consider how well the theoretical ideas developed in this chapter stand the test of research and application.

II

STRESS, SOCIAL SUPPORT, AND HEALTH: EMPIRICAL EVIDENCE

3

EFFECTS OF SOCIAL SUPPORT ON STRESS AND HEALTH I: OF MICE, MEN, AND WOMEN

Chapter 2 has shown that both scientists and laypersons view social support as a potential means of reducing stress, improving health, and buffering people against the deleterious effects of stress on health. But the discussion thus far is largely speculation as to how support *could* or *might* have such effects. What is the evidence that social support actually has such effects, especially with respect to occupational stress and health? And what patterns are evident in the available data that are relevant to those concerned with implementing applied programs to reduce occupational stress and improve occupational health? These questions will be addressed in this chapter and the next.

Everyday experience suggests that it is better to have social support than not to have it. Research helps to confirm popular intuitions here, which is no small achievement as people's intuitions are often wrong. Everyday experience does not, however, provide a sufficient basis for knowing what types of support from what types of persons have what kinds of effects on what kinds of stress and health outcomes. Answers to these questions are necessary if applied programs to reduce stress and improve health by enhancing social support are to be effective and efficient. These answers must come from existing and future research on social support.

A large and growing body of research on the effects of social support and stress on health in both work and nonwork settings exists, yet such knowledge is still often fragmentary and unsystematic. Most of the individual studies or even sets of related studies discussed in the following pages are less than optimal in some way. The conceptual and operational definitions of support (or stress) are not usually comparable across studies, and the populations and study designs also vary greatly. What is impressive, nevertheless, is the consistency of the finding that social support can reduce stress, improve health, and buffer the impact of stress on health. Such findings emerge from studies of both animals and humans, studies done both at a single point in time and over time, and studies of many social settings and populations. Negative results are rare. This accumulating body of data both confirms our belief in the potential importance of social support and suggests specific strategies for enhancing social support in order to reduce stress and improve health.

Chapter 3 will first briefly review some of the evidence on the effects of social support from studies of animals and of humans in nonwork settings. Chapter 4 will then be devoted to a more extensive review of the effects of social support on occupational stress, health, and the relation between them in working populations. Chapter 4 culminates with a summary of key findings that are especially relevant to potential efforts at applying this knowledge of social support.

A NONHUMAN BASELINE

Much understanding of the impact of psychosocial stress on the health of human beings has come from research on stress and health in animals. The work of the physiologist Hans Selye (1976), considered by many the father of modern stress research, is perhaps the prime example (see McLean, 1979, Ch. 2, especially pp. 33–35). Similarly, the potential importance of social support in reducing stress, improving health, and especially buffering the impact of stress on health has been clearly observed in animals. Over 20 years ago Bovard (1959) noted a convergence of observations regarding the supportive effects of social groups on both soldiers and civilians under stress during World War II with findings from animal research on stress:

> *Liddell (1950) found that a young goat isolated in an experimental chamber and subjected to a monotonous conditioning stimulus will develop traumatic signs of experimental neurosis, while its twin in an adjoining chamber and subjected to the same stimulus, but with the mother goat present, will not. . . . A recent experiment by Conger, Sawrey and Turrell (1957) has shown that rats in a chronic conflict situation alone had significantly greater resultant ulceration than animals tested together.*

Cassel (1976, p. 113) cites these same studies and also notes that

> *Henry (1969) has been able to produce persistent hypertension in mice by placing the animals in boxes all linked to a common feeding place, thus developing a state of territorial conflict. Hypertension only occurred, however, when the mice were "strangers." Populating the system with litter mates did not produce these effects.*

More recently, Nerem (1980) and associates have found that if rabbits on a high-fat diet are cuddled, fondled, and talked to (given social support, I would argue) by their handler, they somehow are protected against arteriosclerotic heart disease, while rabbits on the same diet who were not given such special attention generally succumbed to heart disease.

Thus, adverse physical effects of potentially stressful conditions on animals are greatly reduced or even eliminated if the animals are exposed to these conditions in the presence of other familiar animals (or even supportive humans). Other animals facing these same situations alone or in the company of "strangers" (animals of the same species but with no prior relationship or acquaintance) develop serious psychological and physical disorders. Researchers do not know, and may never be able to know, exactly how and why the presence of other familiar animals buffers goats, rats, and mice against the effects of stress. But it appears that something like social support is operating here. Animals, like humans, appear to respond differently to stimuli or situations depending on how threatening they perceive them to be (Mason,

1975). The degree of threat seems to depend on the familiarity of the situation, the suddenness with which new stimuli are introduced, and the degree to which sources of aid or protection are present. Other animals may make the setting more familiar and offer promise of protection. Supportive animals or humans may also directly alter neuroendocrine processes. While researchers can only speculate about what goes through the minds of animals in such situations, they can probe the human mind more directly.

HUMAN BEINGS IN THE LABORATORY

Social psychological experiments on humans have also documented beneficial effects of other people in buffering the impact of stress. These effects are often similar to those in the animal studies, sometimes different. In all cases they extend our understanding of social support.

Back and Bogdonoff (1967) performed a series of experiments in which subjects had a catheter inserted into a blood vessel in their arm and left there for the course of the experimental session. Blood was drawn periodically while subjects performed a seemingly simple perceptual judgment task. This task involved matching one of three stimuli projected on a screen with a standard stimulus, either before or after receiving feedback of other subjects' judgments. These other judgments, however, were often rigged to be contrary to what was obviously and objectively the correct response. Both the drawing of blood and the judgment task were stressful for most subjects, and this was reflected in marked rises in free fatty acids in the blood, an indicator of physiological arousal. However, this arousal was significantly lower among subjects who came with a group of friends than among subjects who came individually. Some subjects who came individually were given a personality test in small groups one day, and then came with the same group to the experimental session. Even this brief previous interaction lowered subjects' arousal in the experimental session, although not as much as the presence of friends.

Thus, humans, like animals, are somehow protected against the effects of experimental stressors by the presence of familiar persons—friends or, to a lesser degree, other subjects with whom they have had one previous interaction. These others appear to provide emotional and appraisal support. That is, being in the presence of familiar others may be reassuring because they explicitly or implicitly convey the feeling that someone is there who knows and cares about us. Further, seeing these others remaining calm in the situation, allows us to appraise the situation as less threatening.

A set of experiments by Schachter (1959) demonstrated that, given the opportunity, people confronting a stress-inducing experience actually seek social support. Experimental subjects who thought they were going to experience a series of painful electric shocks and subjects who anticipated experiencing only very mild electrical stimuli were given the choice of waiting alone or with other subjects for a brief period while final preparations for the experiment were made. Schachter found that the greater the anticipated pain, the more subjects chose to wait with others. He interpreted this pattern as reflecting heightened needs for reassurance, distraction, information, and social comparison (analogous to emotional, instrumental, informational, and appraisal support) among subjects experiencing greater stress.[1]

Sarnoff and Zimbardo (1961) largely replicated Schachter's results, but also demonstrated that people do not always seek social affiliation or support in the face of emotional arousal. They distinguished between *fear* of an inherently dangerous, painful external object (the emotion aroused in Schachter's study and their replication of Schachter) and *anxiety,* which they defined as emotional arousal that has no clear source or stems from objectively harmless objects. In addition to replicating Schachter's fear arousal conditions (anticipation of electric shock), their study also included an anxiety arousal condition in which male college students were led to anticipate having to suck on a variety of nipples, baby bottles, and pacifiers. As expected, based on individual differences in the symbolic significance of the nipple stimuli (see Fig. 2.3), there was more variation in emotional arousal in the anxiety condition than in the fear condition. More importantly, high levels

---

1. Schachter's studies also showed that some individuals felt a greater need than others for affiliation or support in the face of stress. Specifically, he hypothesized and found that first-born children showed both higher anxiety and desire to wait with others in his experiments than did later born children. There are a number of possible reasons for these findings. Note that the need for support varies across individuals as well as across situations.

of emotional arousal in the anxiety condition led to a slight preference for waiting alone rather than with other subjects (whereas almost all subjects in the fear condition chose to wait together).

Sarnoff and Zimbardo (1961, p. 357) suggest that if people are emotionally aroused for reasons that are largely idiosyncratic or not socially acceptable, they fear many kinds of social contact and even support will only increase their anxiety:

> *Because the anxious person tends to be aware of the element of* inappropriateness *in his feelings, he is loath to communicate his anxieties to others. To avoid being ridiculed or censured, he conceals anxiety aroused by stimuli which he guesses do not have a similar effect upon others, and which, he feels, ought not to upset him. Thus, when anxiety is aroused, a person should tend to seek isolation from others. On the other hand, when fear is aroused and he is unable to flee from the threatening object, he welcomes the opportunity to affiliate.*

These ideas received confirmation in a field study by Mechanic (1962) of graduate students preparing for preliminary examinations. Students who were more isolated and less often studied with other graduate students experienced less stress and anxiety in preparing for the exam. It appears that frequent contact with other students only served to make students more aware of all that they did not know and hence heightened rather than reduced their feelings of uncertainty and inadequacy. These findings reinforce the suggestions in Chapter 2 that social support—especially informational and appraisal support—may sometimes exacerbate preceptions of stress and at least short-term responses indicative of personal strain (e.g., anxiety, smoking, heightened blood pressure, insomnia).[2]

The longer-term implications are less clear. Information on whether Mechanic's more isolated students actually performed better or worse is unavailable. Elsewhere, Mechanic (1970) has argued, quite correctly, that what are sometimes called defensive behaviors, which might include avoidance of some forms of social

---

2. Mechanic also found that support from spouses, which was largely emotional support ("No matter how it goes, I'm still with you"), reduced the anxieties and tensions involved in preparing for the exams.

contact and support, can facilitate long-term coping and adaptation by keeping levels of anxiety at manageable levels. On the other hand, anxiety and other responses indicative of personal distress are important signals and motivations for adaptive behavior. Longer-term individual and social growth and development often come at the price of short-term disruption and distress. Thus, when people avoid persons who give them informational or appraisal support that is tension-producing, it may be adaptive in the short run, but not in the long run.

In sum, experimental and field studies of humans also suggest that social support can protect people against short-term aversive effects of stressful situations and that people often seek social support in such situations. However, under some circumstances, support, especially informational and appraisal support, may increase short-term anxiety or other stress responses. The effects of emotional support appear more generally benign. Understanding the interplay between emotional support and other forms of support, in terms of both their short- and long-term consequences, is a major challenge for future basic and applied work on social support. These issues will be discussed further, especially in Chapters 5 and 6.

FIELD EXPERIMENTS

Quasi experiments in naturally occurring stressful situations have further demonstrated the role of social support in buffering the impact of stress on health. Perhaps the single most widely cited study in the literature on social support is that of Nuckolls et al. (1972). The study actually dealt with psychosocial assets, of which social support was a major, but not the only, component. This study of 170 army wives examined the role of potentially stressful life changes occurring prior to and during pregnancy in producing medical complications during pregnancy, and the ability of psychosocial assets to mitigate such effects.[3] Nuckolls et al. found that

3. The measure of life change was essentially that of Holmes and Rahe (1967) that is described in detail by McLean (1979, pp. 65–68). The measure asks about over 40 events or changes ranging from the death of family and friends or separation and divorce to taking a vacation. In arriving at a total score, events are weighted by the amount of readjustment they require.

women who had low levels of life changes *or* high levels of psychosocial assets experienced a moderate rate (33 percent to 49 percent) of complications of all types, but almost all women (91 percent) with high levels of life change *and* low levels of assets experienced complications. Thus, life changes increased complications, but only if psychosocial assets, including social support, were low. Women with high psychosocial assets, including social support, were protected or buffered against the adverse health effects of a high rate of life changes. These assets, however, were not especially beneficial to women who experienced low levels of life change.

A study by Raphael (1977) demonstrated that providing social support to recently widowed women in the form of a few hours of supportive and nondirective psychiatric counseling markedly lessened the rate of impairing physical and mental illnesses experienced by these women. Sixty-four recent widows judged to be at high risk of mental and physical disorders were randomly assigned to an intervention group or a control group that received no intervention. Those in the intervention group received a few hours of supportive, nondirective individual counseling in their own homes. The "goals of the intervention were the promotion of normal grieving—expression of bereavement affects and the accomplishment of a significant degree of mourning —review of positive and negative aspects of the lost relationship, and gradual going over and giving up" (Raphael, 1977, p. 1491). All counseling terminated within three months of the husband's death.

Thirteen months after their husbands' deaths both intervention and control groups reported via questionnaire on health changes during widowhood. Expert physicians judged that any widow experiencing 16 or more changes had suffered a "major health impairment." Whereas almost 60 percent of the control group had experienced a major health impairment since their husbands' deaths, less than 25 percent of the intervention group exhibited such adverse outcomes.

Thus, experimental and quasi-experimental field studies further indicate that support from both professional and lay sources can buffer people against potential deleterious health effects of naturally occurring psychosocial stress, though the events and health outcomes in question are generally short term and of moderate intensity. More consequential health outcomes and more chronic life stresses have been examined in a long series of nonexperimental studies.

NONEXPERIMENTAL RESEARCH

One of the most frequently reported findings in social research for almost a century is that socially unattached, isolated or "marginal" individuals are characterized by lower levels of physical and mental health and higher rates of mortality than people who are more socially attached and less isolated or marginal. One of the seminal works of modern sociology was a study of the social correlates of suicide conducted by a French sociologist, Emile Durkheim, around the turn of the century. Durkheim showed that Protestants and unmarried persons committed suicide more often than Catholics and married persons, respectively, and attributed these results to the lesser degree of social integration and attachments among the former than the latter. Subsequent research has consistently found a lower level of physical and mental health among socially isolated and marginal persons (Cassel, 1970; Jaco, 1970; Kaplan et al., 1977). The poorer health of unmarried persons, compared to the married, is especially well documented (Bradburn, 1969, Ch. 9; Chen and Cobb, 1960; Gove, 1972, 1973; Pearlin and Johnson, 1977).

Most of this research is purely cross-sectional in nature documenting a correlation between social ties and physical and mental health. But such correlations could be produced either by effects of social relationships on health or by the effects of health on social relationships. Perhaps it is not that lack of social ties and support impairs health, but rather that people with impaired physical and/or mental health have difficulty forming social ties and obtaining social support. Fortunately, two other kinds of studies suggest that much, if not all, of the causal flow is from social relationships to health rather than vice versa.

First, a variety of studies have looked at the short-term effects of the *loss* of social attachments and relationships on health. Almost invariably these studies show that the loss of important family and friendship ties through death or separation is followed by a marked rise in physical and mental health problems. The problems of widows in the study by Raphael (1977) just discussed provide one example of these effects. Kaplan et al. (1977) cite a number of other relevant findings. Much of the voluminous literature on the deleterious effects of life events and life changes on health can be interpreted as evidence of the effect of loss of social ties and supports. Four of the five most important events in the Holmes and Rahe (1967) Schedule of Recent Events are essentially losses of social ties and supports (that is, death of spouse, divorce, marital separation, death of a close family member). These events are not uncommon and undoubtedly constitute a major, perhaps the major, component of high scores on the Holmes and Rahe measure.[4]

Second, there are occasional long-term, longitudinal studies of the impact of social supports on health, most notably the work of Berkman and Syme (1979). These researchers analyzed data gathered between 1965 and 1974 on 2,229 men and 2,496 women, aged 30 to 69 in 1965 and randomly sampled from the population of Alameda County, California. They assessed whether the presence or absence of four kinds of social ties in 1965—marriage, contacts with friends, church membership, and informal and formal group associations—affected the likelihood of the person dying over the next nine years. People low or lacking in each type of social tie were from 30 percent to 300 percent more likely to die than those who had each type of relationship. Generally, these trends hold for both sexes and at all age levels, although marriage had the strongest protective effect for men while contacts with friends were most protective for women.

Each of the four sources of contact predicted mortality independently of the other three, but the more intimate ties of marriage and friendship were stronger predictors than were ties of church and group membership. Berkman and Syme also combined the four sources of social contacts into a Social Network Index, weighting the more intimate and important ties more heavily.[5] Fig. 3.1 (from Berkman and Syme, 1979) shows that people with the least connections were 1.8 to 4.6 times more likely to die than persons with the most connections although the numbers in the

4. I am indebted to John R. P. French Jr. for the interpretation of life change as loss of social support.

5. Details on the measures and other aspects of the methods of this study are presented in Berkman (1977).

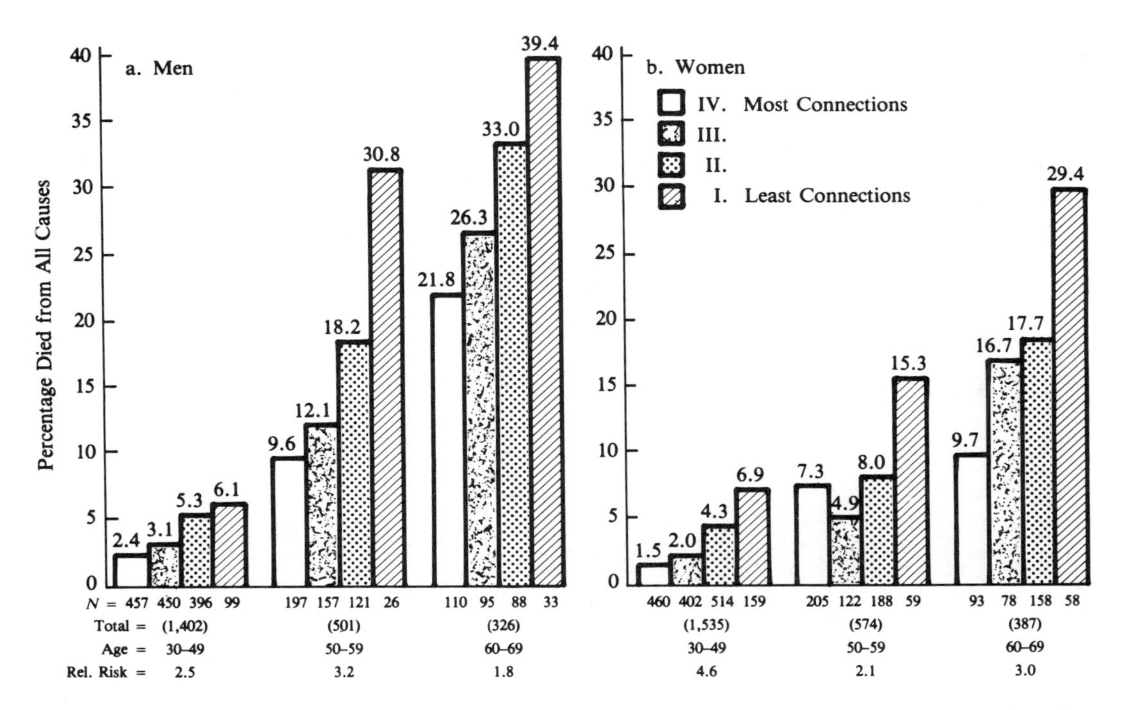

FIG. 3.1 *Age and sex-specific mortality rates from all causes per 100 for social network index, Human Population Laboratory Study of Alameda County, 1965 to 1974 (Berkman and Syme, 1979, Fig. 1).*

least connections group are fairly small (see "Rel. risk" or relative risk in Fig. 3.1). These differences remain strong and significant even after controls for self-reported health status at the beginning of the study, socioeconomic status, health practices (for example, smoking, drinking, obesity, physical activity) and use of preventive health services (annual medical and dental checkups), although a portion of the overall advantage of the socially connected people stemmed from their greater use of both disease-preventive health practices and health services.

The Berkman and Syme data are striking and compelling, because of both their strengths and weaknesses. They document a striking predictive impact of social ties on that most final of indicators of ill health—death. These data thus provide a strong argument that previous cross-sectional correlations between social isolation and health reflect a causal impact of social ties on disease. However, Berkman and Syme assessed only the presence of social relationships, not the quality or content of those relationships. About 60 percent of their sample fell into the two most connected categories of their Social Network Index and showed only slight mortality differences. The big differences were between these two groups and the less-connected people, particularly the most most isolated. Attention to the quality and content of relationships would help to differentiate among people with social ties and would very likely show an even stronger difference between people with truly supportive social networks and those with little or no supportive relationships. Further, Berkman and Syme made no attempt to assess exposure to psychosocial stress or other health hazards, against which the effects of social ties and supports may be especially protective. That is, they did not test directly for the buffering effects of support. Thus, there is reason to believe that a study encompassing more careful measurement of both social support and stress would have produced even stronger evidence of the predictive power of social support in relation to mortality.

CONCLUSION

A wide range of evidence from laboratory and field studies of both animals and people indicates that the presence of social relationships, especially if they are characterized by supportive behavior, can reduce the experience of stress, improve health and/or buffer the impact of stress on health. The review of studies has been necessarily selective. Epley (1974) provides a more extensive review of the evidence from animal studies, while Cassel (1976), Cobb (1976), and Kaplan et al. (1977) do the same for studies of humans. Some skepticism and reservations have been voiced, however, regarding the strength of the evidence that social support reduces stress, improves health, and especially buffers the impact of stress on health. Two questions seem especially frequent and important.

The issue of causality

Some, notably Heller (1979), question whether social support causes good health or good health makes a person more likely to experience social support. Without carefully controlled experiments which manipulate the levels of social support experienced by people over long periods of time—experiments that are impractical and/or unethical—researchers can never definitively answer this objection. Without similarly impractical and unethical human experiments, however, scientists can never establish definitively that cigarette smoking causes lung cancer and heart disease, as the tobacco companies frequently point out. Nevertheless, most people, and certainly most biomedical scientists, have all the evidence they need, short of these impossible experiments, to say that smoking is hazardous to health. Likewise, I think the combination of available laboratory experiments on animals and cross-sectional, quasi-experimental, and longitudinal studies on humans make an increasingly compelling case for the causal impact of social support on stress and health, and the relation between them. A skeptic may still see the glass as half empty; I think it is at least half full and becoming fuller all the time.

The issue of buffering

Pinneau (1975, 1976) and others (Pearlin and Schooler, 1978) have raised a more limited and more complicated reservation regarding the existing evidence. They argue that social support may indeed reduce stress and improve health, but there is little hard evidence for the much touted buffering effects of support, that is, the ability of support to diminish and even eliminate the deleterious effects of stress just as a sunscreen diminishes or eliminates the deleterious effects of the sun or an effective influenza vaccine diminishes or eliminates the deleterious effects of certain strains of influenza virus. As noted in Chapters 1 and 2, the importance of social support does not rest exclusively on its purported buffering effects, but these effects are among its most unique, intriguing, and practically important qualities. Thus the reservation must be seriously considered.

Again, although there is some basis for this reservation, it has, I think, been overdrawn. The careful reader may have noted that all of the experimental and quasi-experimental studies reviewed here explicitly or implicitly involved a buffering effect of social support on the relationship between stress and health, whereas most of the nonexperimental studies documented, in the terms of Chapter 2, main effects of social ties and supports on health. In some of the experimental studies, most notably Nuckolls et al. (1972), there is an explicit comparison of the impact of social supports among persons experiencing high stress versus low stress. In most experiments, however, there is no low-stress condition, but rather social ties and supports are shown to make a difference under stressful conditions. A reasonable inference is that support would make little or no difference under low-stress conditions, which most likely would simply not produce the responses of interest. The implication is that support is especially protective of health when people are under stress—the essential idea of the buffering. In contrast, the nonexperimental studies, which are largely cross-sectional, tend to merely correlate social support with health, taking no account of stress levels. Essentially, such studies document main effects of stress on health.

These nonexperimental results seem to suggest that support has main effects on stress and health, but its much touted buffering effects do not exist. However, it is important to note: (1) that many studies (e.g., Berkman and Syme) did not explicitly test for buffering effects, (2) that reanalysis of at least one of the presumed cases of negative evidence (Pinneau, 1975, 1976) has produced positive evidence of buffering effects (see LaRocco, House, and French, 1980; Chapter 4), and (3) that cross-sectional nonexperimental studies are much weaker than experimental designs (and, to a lesser degree, nonexperimental longitudinal studies) for testing buffering effects.

The technical reasons for this last point are developed more fully in Appendix B. Briefly, if support buffers the effect of an objective stressor or perceived stress on health (or any other relationship in Fig. 2.3), this buffering occurs over time. For example, social support may help a person cope with and hence eliminate an objective stressor that would otherwise adversely affect health, or support may buffer the impact of perceived job stress on health by causing it to be interpreted as less stressful. At the time the coping or reinterpretation involved in the buffering process are occurring, researchers would observe differences in the impact of the objective stressor or perceived stress on health between persons with high and low social support. But once the buffering process is complete observers would merely note lower stress levels or better health among persons with greater versus lesser social support—what are here termed main effects of support. In contrast, experimental designs that allow researchers to observe the buffering process directly by establishing conditions of high and low support and then exposing persons or animals in each condition to different levels of stress are better able to document buffering effects.

Retrospect and prospect

In several ways the nonoccupational data reviewed in this chapter provide a necessary background and context for the evidence considered in Chapter 4 on social support and work stress. First, although the data in this chapter do not deal specifically with work, they clearly show that in many social situations and for many individuals social support can and does reduce the experience of stress, improve health, and/or buffer the impact of stress on health. This fact makes it reasonable to hope and expect that social support can have similar salutary effects in work settings as well. Second, the boundary between work and nonwork is highly flexible and permeable. Nonwork stress and support can affect workers' functioning and health on the job while work-related stress and support can affect people's functioning and health outside of work. People often carry their problems from home with them to work, while their occupational stresses often go home with them in one form or another. Similarly, support from family, friends, and relatives can and sometimes does help people to adapt to occupational stress, and their social supports at work may similarly help them deal with stress outside of work. Thus, neither researchers nor practitioners can adequately understand and deal with work stress and support without also considering nonwork stresses and support (and vice versa).

Finally, much of the research on work stress and social support has both profited from, and sought to fill in gaps in, prior research on stress and support more generally. As LaRocco et al. (1980, p. 214) observe, for both scientific and practical reasons, "it is time to stop simply 'proving' that social support is related to stress and strain and begin to consider the mediating factors or mechanisms through which social support functions." This consideration requires that researchers be able to answer more concretely and specifically the central question regarding social support posed in Chapter 2: *who* gives *what* to *whom* regarding *which problems?* Only this kind of more concrete and specific knowledge can serve as the foundation for practical application and intervention. And researchers have begun to generate just this kind of knowledge in the area of work stress and social support.

4

EFFECTS OF SOCIAL SUPPORT ON STRESS AND HEALTH II: THE WORK SETTING

This chapter will focus on an intensive review of several recent studies indicating that certain kinds of social support from certain kinds of people can reduce certain kinds of occupational stress, improve certain health indicators, and buffer certain relationships between stress and health. Even before these recent studies of work stress and social support, a good deal of evidence from organizational psychology and sociology strongly suggested that social support could play a major role in alleviating occupational stress and health problems. This chapter begins with a review of some of this early evidence and then turns to the more recent focused studies. The chapter concludes with a summary of implications of existing research for organizational practice—implications that are developed in Chapters 5 and 6.

Social support does not emerge as a panacea for all occupational stress and health problems. But it is clear that the right kind of support from the right kind of people can be of significant value in reducing occupational stress, improving health, and buffering the impact of stress on health. Further, as Chapters 5 and 6 will argue, we can probably do more in the work setting than elsewhere to enhance social support, hence reducing stress and improving health. It is difficult, if not impossible, to insure that most people have a supportive spouse, as important as we know this to be from the data in Chapter 3. It is more possible to insure that almost all employees in work organizations have access to at least one supportive person at work. This chapter provides the foundation for such efforts.

SOCIAL SUPPORT AT WORK: PRELIMINARY EVIDENCE

Implicitly, if not explicitly, much of organizational sociology and psychology over the past 20 to 40 years can be interpreted as indicating that the experience of social support at work can reduce occupational stress and/or improve occupational physical and mental health. The human relations tradition of organizational research (Mayo, 1933; Likert, 1961, 1967) has long emphasized that supportive behavior by work supervisors can improve both the morale and productivity of workers and reduce many forms of organizational stress. Kahn and Katz (1960) provided an early review of evidence for such positive effects of supportive supervision from both cross-sectional survey studies and actual experiments where organizational support from coworkers also was found to have stress-reducing effects. Seashore (1954) found that as work group cohesiveness increased, anxiety over work-related matters decreased, and Likert (1961) and Kahn, Wolfe, Quinn, Snoek, and Rosenthal (1964) reported similar results.[1]

More recently, there has been some shift of interest away from purely interpersonal aspects of the work environment toward greater emphasis on the impact of the social and technological structure of work on both organizations and individuals. Attempts at planned organizational change have sought to modify the organizational structure itself, as well as interpersonal relations within that structure, on the grounds that such changes produce more profound and enduring effects. In reviewing the literature on planned organizational change Katz and Kahn (1978, Chapters 19 and 20) emphasize that larger-scale or systemic changes have had the great-

1. Note that, as in Chapter 3, the effects of social support, especially informational and appraisal support, are not always salutary for individuals or organizations, at least in the short run. For example, Seashore's research (1954) and earlier work, including the classic Hawthorne studies, suggest that cohesive work groups can reinforce norms promoting job dissatisfaction and restriction of productivity.

est impact on the quality of working life. They note especially (1) the experiments by London's Tavistock Institute with the socio-technical system of British coal and Indian textile mills and (2) experiments in Norway and Sweden with the development of semi-autonomous work groups. The major thrust of both the Tavistock and Scandinavian efforts has been to reverse a trend toward fractionation and specialization in industrial jobs by giving a small group of workers responsibility for a set of tasks and allowing them to decide how these tasks will be organized or divided among them. The primary focus is the structure of the work process rather than quality of interpersonal relations. However, the impact on interpersonal relations may be the more important outcome.

In the well-known Volvo automobile assembly plant at Kalmar, Sweden, the assembly process was designed so that instead of each worker performing a single task along a continuous conveyor line, groups work together in a bay on a cluster of assembly tasks (for example, the installation of the motor). Cars move from bay to bay but within each bay workers may cooperate and rotate in performing a series of tasks each of which would be the province of a single worker on a traditional assembly line. The Tavistock studies of British coal mines involved similar changes, though in this case the change represented an adaptation of more traditional mining methods to a new technology. Traditionally groups of coal miners had responsibility for mining a face of coal, and all miners shared in all aspects of the operation—cutting, drilling, blasting, loading, and extending roof supports. With the introduction of special heavy machines for each of these tasks, each miner became a specialist in the operation of one machine and often each shift of workers specialized in only one or two operations. The result was decreased worker satisfaction, higher absenteeism, and lower productivity—problems also associated with traditional automobile assembly line technology. In successfully remedying these problems, the Tavistock researchers essentially imposed the old social organization of mining on the new technology. Workers on a given face were organized into teams that were again responsible for carrying out all mining operations, with the organization and distribution of jobs decided by the team. Thus, the team could again experience the full cycle of mining and each individual could perform many jobs.

These experiments have generally resulted in lower levels of absenteeism, higher productivity, and reduced levels of a range of perceived occupational stresses (for example, job dissatisfaction and job-related tension), though the evidence on the results of the recent Scandinavian experiments is somewhat fragmentary at this point. The usual explanation for these effects is primarily that workers experience a sense of job enlargement and enrichment, and secondarily that they form stronger work group ties. Reflecting on these studies I was struck by the possibility that the change in social relations may be the major explanation of these effects, and it is clearly the major way in which these experiments differ from conventional experiments in job redesign, which focus on individual jobs and have generally had less profound effects. The Tavistock and Scandinavian experiments not only improve the experience of individual workers but also involve workers in *cooperative, task-related, group* interaction.[2] These conditions (discussed in Chapter 5), are probably optimal for workers giving and receiving all types of social support. Indeed, I would hypothesize that one of the most important effects of such experiments in organizational change is to enhance levels of social support from coworkers and supervisors, and that this enhanced social support provides a major explanation of the stress-reducing and health-enhancing impact of these experiments. If I am right, these studies also provide evidence of a causal impact of changes in social support on levels of occupational stress and health.

In sum, a good deal of basic and applied research in organizational psychology and sociology suggests that social support can reduce occupational stress and improve worker health. However, none of this research was explicitly concerned with social support. It is only in the past decade or so that research has focused directly on the role of social support in reducing occupational stress, enhancing health, and buffering the impact of occupational stress on health. Three such studies will be detailed here.

---

2. This process of interaction began in these cases with participation in the actual planning of organizational change. Such participation in decision making about change also has a variety of positive effects (see French, Israel, and Aas, 1960), and it may be crucial to the overall impact of experiments in planned organizational change—not all of which have proved as successful as the cases cited here.

SOCIAL SUPPORT, JOB CHANGE, AND HEALTH

Changing jobs, voluntarily or involuntarily, is an increasingly prevalent feature of the modern world of work. Economic and technological forces confront many workers, in white collar as well as blue collar jobs, with the loss of their jobs and the need to find new jobs. Slightly less disruptive than job loss is the frequent job mobility experienced by workers in many occupations and industries. Even where workers do not formally change their jobs, the nature of their jobs may change radically around them as a result of technological and organizational change. Thus, job change is a major potential source of occupational stress in our society.

The role of social support in protecting people against the adverse effects of job change and job loss was first noted in studies of unemployed workers during the Great Depression. Liem and Liem (1979) summarize how these studies documented:

> *...the despair, isolation, and resignation accompanying the separation of the individual and the family from the workplace, and the quality of interaction within the family that ameliorates these effects. Komarovsky (1940), for example, noted that the nature of the preunemployment relationship between a worker and his spouse was a critical determinant of the degree of deterioration of family functioning that took place throughout the course of unemployment. Egalitarian marital relationships based on love and respect were more likely to promote a continued family stability in the face of prolonged unemployment than were patriarchal relationships of a utilitarian nature.*
>
> *Similar attention is drawn to social relationships as important buffers against the stresses of job loss in Bakke's report (1940) of family conditions that foster optimal coping during extended unemployment. Bakke found that the ability and willingness of other family members to assume the provider role following unemployment of the primary breadwinner served to decrease the disruption of family stability. He further observed that extrafamilial relationships were important for sustaining family stability because they provided emotional support and encouragement, financial assistance, job leads, and social pressure to maintain the integrity of the family. Each of these is obviously an example of an emotional*

> *or instrumental support. Jahoda, Lazarsfeld, and Zeisel (1971) suggest that differences in family and social relationships, together with differences in financial resources and individual skills and abilities, were responsible for the varied responses to job loss they observed among unemployed workers in Marienthal. Furthermore, they note, as did Bakke, that over time unemployment stress may tax the very relationships that earlier served to moderate its negative effects.*
>
> (LIEM AND LIEM, 1979, pp. 359–360)

More recently Cobb and Kasl (1977) and Gore (1978) have reported the effects of social support on the levels of stress and health experienced by male industrial workers in two plants over a two-year period during which the plants were closed and the men lost their jobs of longstanding and had to find new jobs. The men were interviewed five times about their occupational, financial, familial, and health status: (1) at *anticipation* or six weeks prior to the scheduled closing but after the men knew of the closing, (2) at *termination* or one month after closing, and (3) *six months,* (4) *twelve months,* and (5) *twenty-four months* after the closing. The study provides quite striking evidence of the effect of social support in reducing the level of perceived stress, improving health, and especially buffering the impact of stress on health. The study is sufficiently complex, however, that the interpretation of the data presented here is somewhat different from that presented in the major report of the study (Cobb and Kasl, 1977).

The important stressor in this study turned out not to be loss of a job per se, but how long the worker was unemployed. Further, the impact of job loss and unemployment was evident well before the plant actually closed. That is, the men began to experience stress at the time they first heard that the plant would be closed, which was two years prior to the final closing (Slote, 1969). Further, their reactions to the planned closing, evident in the anticipation phase of the research, seemed to reflect an understanding of the difficulties they would experience in finding reemployment.

About 50 percent of the workers were reemployed by the time of the first interview after the plant closing (the so-called termination interview). The mean number of weeks unemployed was about 10 for the first year and about 15.5 over the full 24 months of the study. Thus, the majority of workers experienced a relatively brief period of unemployment, but some experienced a great deal. On almost all measures of interest in the study, the greater the length of unemployment the greater the level of perceived stress and poor health experienced. In most analyses, the measure of severity of unemployment was a dichotomous variable indicating whether the worker was or was not unemployed for 10 percent or more of the first year after the plant closing. Workers who experienced more unemployment score higher on indices of perceived stress, strain, and poor health at all phases of the study including anticipation, suggesting that most workers had a good idea before they lost their job of how soon they were likely to be reemployed. Some differences between those who experienced more or less unemployment tend to diminish toward the end of the study when almost all workers were reemployed.

Unemployment experience generally had adverse effects, however, only if workers had low social support. That is, high levels of social support tended to protect the workers against the adverse effects of more prolonged unemployment. This buffering effect of support was evident on a wide range of indicators of work-role deprivation and mental health or well being and on selected physiological and physical health variables as well.[3]

Fig. 4.1 presents two representative findings. The left-hand graph shows the trends in a summary measure of work-role deprivation across the five phases of the study for four groups of workers—(1) those with *low* social support (below the median) and *more* unemployment (more than five weeks in the first year), (2) those with *low* support and less unemployment, (3) those with *high* support and *more* unemployment, and (4) those with *high* support and *less* unemployment. The measure of work-role deprivation asked the men how much they felt they were *not* getting many of the benefits and rewards normally derived from work

3. The measure of social support in this study combined thirteen questions: six on the perceived supportiveness of wives, one each on the perceived supportiveness of friends and relatives, two on the person's ability to get support from "people around you" and one each on rates of affiliative behavior with family, friends, and relatives. The measure taps mainly emotional support. Because some items were not asked in all phases of the study and those that were varied little across phases, a single measure of support was computed by averaging across all questions and phases. The supportiveness of wives was clearly the most important component of this measure.

(for example, physical activity, security, acquiring and using skills, respect, authority), except money which was asked about in a separate measure of economic deprivation. The scores on work-role deprivation (and on depression) are expressed as deviations from the score of a control group of workers who were stably employed throughout the study in similar plants.

With one slight exception the work-role deprivation scores of *men who experienced less unemployment or had high social support* fluctuate right around the mean of the control group (the "0" point in Fig. 4.1). That is, these men *were no worse off than workers who remained stably employed* over the course of the study. The group with high support but more unemployment experienced a momentarily elevated level of deprivation at the termination interview, probably a realistic response to still being unemployed at that point, but quickly dropped back to the range of the less unemployment groups. In marked contrast, *men who experienced more unemployment but had low social supports responded with much higher levels of deprivation throughout the study*, though they slowly fell back toward the other groups as the study progressed and they found stable reemployment.

A similar pattern of results was obtained for all of the component measures of the work-role deprivation index for which Cobb and Kasl (1977) present data. A separate measure of economic deprivation shows the same pattern except that the low support, more unemployment group does not differ from the others on economic deprivation until the termination interview when their income actually dropped for the first time. Generally, this group came closer to equalling the other groups at 24 months in measures of extrinsic deprivation (economic, security, respect) than on intrinsic deprivation (using skills). In general, workers were able to find new jobs that gave them income and respect, but these were often jobs that bore little relation to the skills they had acquired in their former jobs. In sum, social support clearly buffered workers against the effect of unemployment on work-role deprivations, but considered separately, neither support nor unemployment had clear effects on work-role deprivations.

Similar results were also obtained for more general indicators of mental health and well being. The right-hand graph of Fig. 4.1 presents the relevant data for depression. Again the low support, more unemployment group stands out from the others, although

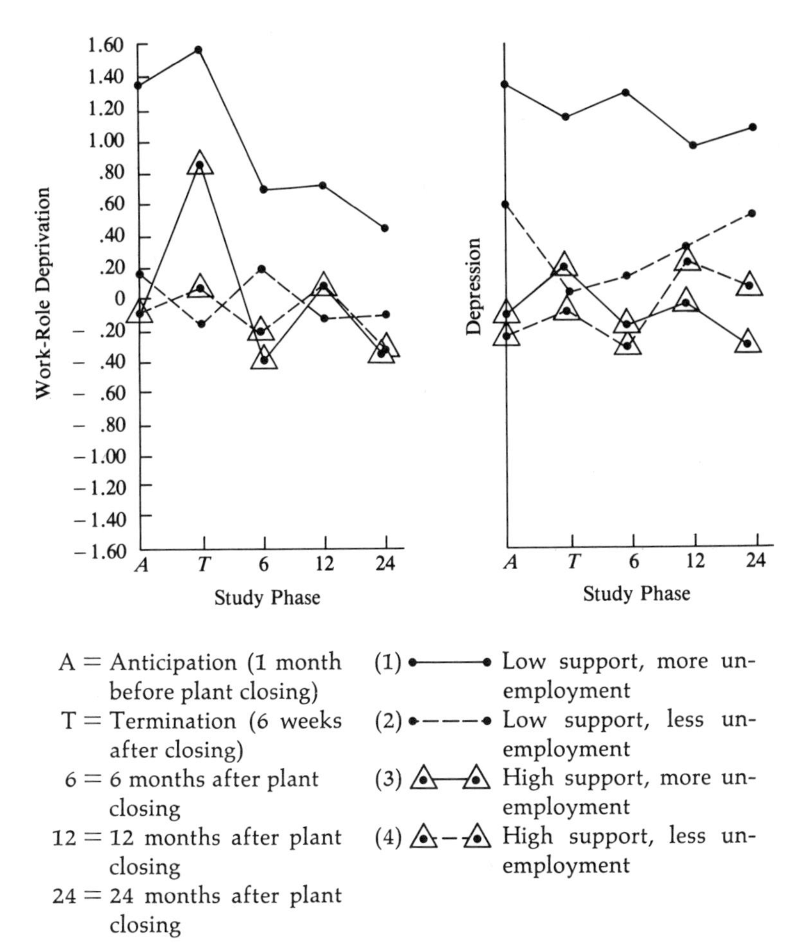

NOTE: All scores are expressed as scored deviations from the scores of a control group of workers who were stably employed throughout the study phases in a similar plant.

FIG. 4.1 *The joint impact of social support and unemployment on work-role deprivation and depression (Graphs created from data presented in Tables 4.8 and 4.15 of Cobb and Kasl, 1977).*

the low support, less unemployment group generally manifested slightly more depression than the two high support groups. Again, there is a strong buffering effect of social support, with support also exerting a slight main effect on depression as well. Most important, however, *the negative effects of unemployment on depression are completely eliminated by adequate social support.* These results for depression were essentially replicated on seven of eight other measures of mental health (low self-esteem, anomie, anxiety-tension, psychological symptoms, insomnia, suspicion, and resentment).

The effects of unemployment on physiological variables and disease states were more complex, generally less strong, and buffered by social support in only a couple of cases. Both unemployment and social support had their strongest effects on the incidence of marked symptoms of rheumatoid arthritis—men's reports of swelling in two or more joints. Cobb and Kasl (1977, pp. 140–141) found joint swelling markedly higher among the terminated workers than among the stably employed control group, but *the difference was primarily due to a 41 percent rate among terminees with low social support* (versus 12 percent and 4 percent among terminees with medium and high support, respectively, and 6 percent among controls). Gore (1978) and Cobb and Kasl (1977, p. 101) found that cholesterol levels of the terminees rose between anticipation and termination and then fell during the remainder of the study—both of these trends being most marked for workers with low support and more unemployment, the group that showed the highest levels of work-role deprivation and psychological symptoms in Fig. 4.1.

Some of the men in the Cobb and Kasl study were also interviewed intensively by journalist Alfred Slote, who described their ordeal in a book—*Termination: The Closing at Baker Plant* (1969). Although published before most of the data discussed above were fully analyzed, Slote's observations are remarkably consistent with the scientific results:

> *The difference between success and failure might seem to be in a mixture of luck,* job skills, the kind of wife a man had, *his age, intelligence and emotional make-up.*
>
> (SLOTE, 1969, p. 323, emphasis added)

Most of these factors were ones that facilitated men's becoming stably reemployed relatively quickly. The "kind of wife a man had," however, was crucial if he did not make that transition easily:

If the wives suffered alongside their husbands . . . it is also true that many of them provided the buttressing strength that enabled their men to weather the change. Joe Nadeau who left the plant in a drunken rage on his termination day had a wife who worked, who supported him emotionally, who made decisions for him. Nadeau, who is extremely prone to psychosomatic illnesses and had ulcers and headaches the last three months of the closing, has made it on his new job thanks mainly to a tough and supportive wife.

(SLOTE, 1969, p. 325)

Another worker Slote calls Henry Burns credited his wife with helping him weather the initial termination experience and his subsequent inability and unwillingness to remain on the first new job he found:

"The next morning (after the plant closing) my wife—God bless her, she's the best woman in the world—she says to me, 'We're going to hitch up the trailer and take a trip.' . . . My wife was wonderful. I forgot all about Baker, about everything. I felt good. I was pretty confident about the future. . . . Finally I got a job at Chevrolet Forge as an ash handler . . . I tell you, it was awful. . . . I came home one afternoon sweating all over and right at the corner up there I blacked out. . . . My wife, God bless her, stood by me. I told her I was scared of that place. I was too old for that kind of work. She stood by me when I called the next morning and told them I was quitting."

(SLOTE, 1969, pp. 191–192)

Mrs. Burns reported "simply that there were many hard moments but they tried to share them together . . ." (Slote, 1969, pp. 191–192). She provided mainly emotional support, whereas Joe Nadeau's wife had provided emotional, instrumental, informational, and probably appraisal support as well. Their roles were critical to their husbands' successes in adapting to this stressful experience.

Slote insightfully brings out that wives not only gave support, but also needed to receive it: "The wives suffered alongside their husbands—and in some cases more than their husbands" (Slote, 1969, pp. 324–325). As students of the Depression have observed unemployment is a family experience. Yet, many wives were unable to provide the support their husbands needed and many husbands were unable to provide the support their wives needed. One of the great paradoxes of unemployment and occupational stress, more generally, is that they often tend to weaken, and even destroy, the very sources of support that are most necessary to cope with stress. This paradox is discussed further in Chapter 5.

In sum, both Cobb and Kasl's (1977) scientific study and Slote's qualitative report provide evidence of the ability of social support to buffer workers against the strains caused by unemployment due to plant closings, especially work-role deprivations, psychological distress, and to a lesser degree physical disorders. Note that the measures of support in this study deal almost exclusively with what I have termed emotional support. Support did not have direct effects on the number of weeks of unemployment experienced (Cobb and Kasl, 1977, p. 163; Gore, 1978, p. 160) or on health. Rather, like most other experimental and quasi-experimental studies, this study primarily provides evidence of the buffering effects of social support. A wider range of support effects becomes evident in two cross-sectional studies of work stress, social support, and health.

STRESS, SUPPORT, AND HEALTH IN A FACTORY

My colleagues and I have studied the effects of social support on work stress, health, and the relationship between stress and health in the hourly workforce of a large tire, rubber, chemicals, and plastics manufacturing plant in a small northeastern city. The data presented here derive from a mail questionnaire returned by 1,809 (out of 2,854 for a 70 percent response rate) white male workers in the plant. (There were too few blacks and females for analysis). The measures of stress and health are generally as valid and reliable as any available, although their self-report nature raises complex questions of validity and causal ordering, which are discussed more extensively elsewhere (Wells, 1978, Chapter 1, and House et al., 1979). Overall, the quality of the data, the striking fit of the results with theoretical ideas about support, especially the buffering effects, and the replication of these results in a second study (discussed next) all suggest that the results to a large extent reflect true causal effects of support on stress, health, and the relationship between them.

TABLE 4.1 *Measures of social support used by House and Wells (1978)*

1. How much can each of these people be relied on when *things get tough at work?*

	Not at all	A little	Some-what	Very much	
A. Your immediate supervisor (boss)	0	1	2	3	
B. Other people at work	0	1	2	3	
C. Your wife (or husband)	0	1	2	3	Not Married
D. Your friends and relatives	0	1	2	3	

2. How much is each of the following people *willing to listen to your work-related problems?*

A. Your immediate supervisor (boss)	0	1	2	3	
B. Other people at work	0	1	2	3	
C. Your wife (or husband)	0	1	2	3	Not Married
D. Your friends and relatives	0	1	2	3	

3. How much is each of the following people *helpful to you in getting your job done?*

A. Your immediate supervisor	0	1	2	3
B. Other people at work	0	1	2	3

Please indicate *how true* each of the following statements is of your *immediate supervisor.*

	Not at all true	Not too true	Somewhat true	Very true
7. My supervisor is *competent* in doing (his/her) job.	0	1	2	3
8. My supervisor is very *concerned* about the welfare of those under him.	0	1	2	3
9. My supervisor goes out of his way to *praise* good work	0	1	2	3

The questions on social support used in this study are presented in Table 4.1. The questions aimed to distinguish between two types of support—emotional (questions 1, 2, 8, and 9) and instrumental (questions 3 and 7) from four different sources—work supervisors, coworkers ("others at work"), spouses, and a combined category of friends and relatives. The study succeeded in the latter aim, but not in the former. That is, workers did not distinguish between emotional and instrumental support, tending to rate others similarly on both measures.

Thus, we arrived at a single measure of support from each of the four sources by adding up responses to the appropriate items in Table 4.1 (six items for the supervisor, three on coworkers, and two each on spouse and friends/relatives).[4] These indices should be viewed as measuring primarily emotional support. We expected that work-related sources of support (supervisor or coworker) would have the greatest effects on stress and health, although nonwork sources, especially spouse, would also be consequential.

We examined the effect of perceived social support from these four different sources taken singly and together (that is, a measure of total support simply summing the four sources) on:

a. five health outcomes (angina, ulcers, itch and rash on skin, persistent cough and phlegm, and neurotic symptoms);

b. seven indicators of perceived occupational stress—job satisfaction and occupational self-esteem (lack of either is considered stressful), work load, role conflict, responsibility, conflict between job demands and nonjob concerns such as family life, and quality concern or worry over not being able to do one's job as well as one would like; and

c. all thirty-five possible relationships between the seven stresses and five health outcomes.

House et al. (1979) fully describe the perceived stress and health measures and show that perceived stress is significantly associated with rates of ill health. Table 4.2 shows the Pearson correlations (*r*'s) of the social support measures with each other and with the stress and health outcomes.[5]

---

4. Unmarried persons were scored 0 on the spouse support measure and 1 point was added to the scores of all married persons after analysis showed that unmarried people were worse off in terms of health than married persons reporting the lowest amount of support.

5. Pearson's correlation coefficients (*r*'s) theoretically range from −1.00 to +1.00, with .00 indicating no relationship, −1.00 a perfect negative relation-

TABLE 4.2 *Correlations of social support variables with each other and with health outcomes and work stresses: (N = 1809 White Male Rubber Workers)*

	Supervisor Support	Coworker Support	Wife Support	Friend and Relative Support	Total Support
Social Support					
Supervisor Support	—				
Coworker Support	.30	—			
Wife Support	.11	.29	—		
Friend & Relative Support	.16	.32	.62	—	
Total Support	.78	.62	.58	.64	—
Health Outcomes					
Angina Pectoris	—.04	—	—	—	—
Ulcers	—.06	—	—	—	—.05
Itch and Rash	—.10	—	—	—	—.06
Cough and Phlegm	—.05	—	—	—	—.07
Neurosis	—.07	—	—.06	—	—.10
Perceived Work Stresses					
Job Satisfaction	.38	.22	.06	.14	.36
Occupational Self-Esteem	.23	.12	—	—	.19
Job-Nonjob Conflict	—.23	—.10	—	—	—.17
Role Conflict	—.22	—.10	—	—	—.19
Responsibility	—.13	—.04	—	—	—.11
Quality Concern	—.39	—.11	—	—	—.29
Workload	—.12	—.08	—	—	—.10

NOTE: Data are excerpted from Table 1 of House and Wells (1978). All coefficients are significant ($p < .05$, one-tailed). Nonsignificant coefficients are omitted and indicated by a "—."

The data in Table 4.2 show several important things. With the exception of a strong correlation between friend and relative support and wife support ($r = .62$), there is only a weak to modest re-

---

ship (as X increases, Y decreases), and +1.00 a perfect positive relationship (as X increases, Y increases). Since these health measures are dichotomies, with only 5 percent to 15 percent of workers reporting positive symptoms of each, the maximum r with these variables ranges from .40 to .60, and even a correlation of .05 indicates a doubling of the rate of illness over the full range of the stress or support variable.

lationship between the supportiveness of the different sources. This finding suggests that workers discriminate between people in reporting their supportiveness. Many workers perceive one source as supportive (for example, supervisor) but not another (for example, wife). Supervisor support tends to moderately reduce all forms of perceived work stress (as supervisor support increases, satisfaction and esteem increase and job pressures such as job-nonjob conflict decrease), and manifests a weaker tendency to reduce disease symptoms. These latter effects are largely indirect—supervisor support reduces stress which in turn improves health. Coworker support has a small to moderate main effect of reducing perceived stress, but no effect on health. In contrast, wife support and friend and relative support have only small and isolated effects on work stress and health. Thus, the data in Table 4.2 indicate that work-related sources of support, especially supervisor support, can reduce work stress and, at least indirectly, improve health; but nonwork sources of support appear to have little or no effect on work stress and health. However, this discussion has considered only main effects; and these data reveal nothing about the potential buffering effects of either work or nonwork support.

For each of the 35 possible combinations of a stress variable and a health outcome, House and Wells tested statistically whether each of the four support measures as well as a combined measure of total support from all four sources buffered each stress-health relationship in a manner consistent with Fig. 2.2. The procedure used is embodied in equation (2) of Appendix A, and the methods and results are presented in detail in House and Wells (1978). A relatively nontechnical overview of the results is presented here.

The index of total *support* reflects the cumulative amount of support perceived from all four sources and approximates the global measures of support used in much previous research (Cobb and Kasl, 1977; Nuckolls, et al. 1972). Given the criteria of statistical significance used (the .10 level) and assuming the 35 tests involving total support are independent of each other, by chance alone the effect of stress on health would differ across levels of support about 3 or 4 times (out of 35) and only one or two of these differences would be true buffering effects of the type depicted in Fig. 2.2. In fact, the effect of stress on health differed significantly across levels of support in nine of 35 tests and *all nine of these were almost perfect buffering effects* (see Fig. 2.2 in Chapter 2)—quite striking evidence that social support can buffer the impact of perceived occupational stress on self-reported symptoms of physical and mental ill health.

It is noteworthy, however, that in almost every case where "total support" had a buffering effect, one of the four support measures composing it did also; and the buffering effect of the single source of support was about as large as the buffering effect of total support. Thus, the effects of total support generally reflect the impact of a single significant source of support *rather* than the cumulative effect of a number of sources of support, none of which alone is sufficient to buffer the worker against the impact of stress. Furthermore, some sources of support are clearly more significant than others in this population. Again assuming independent effects,[6] one or two buffering effects of each source of support would be expected by chance alone. In fact, there were only two buffering effects each for coworker support and friend and relative support—essentially chance level results. However, supervisor support buffered nine of 35 stress-health relationships, and wife support buffered eight.

Fig. 4.2 graphically presents several buffering effects of supervisor and wife support, and compares them to the buffering effects of the index of total support on the same relationships. These graphs show that the effects of support on health in these cases are pure buffering effects, as are virtually all of the buffering effects found in this study. That is, support has no additional main effect on health (there are essentially no differences in health between people with high versus low support when perceived stress is low), although Table 4.2 shows that supervisor support and coworker support do appear to reduce perceived stress levels and hence indirectly improve health. Further, these graphs show that the effects of total support in each case are almost totally accounted for by the effects of either supervisor or wife support. This important finding is not evident in research that uses only a global measure of support.

---

6. This assumption is clearly not justified here, but is used as a heuristic device for establishing a lower bound below which any results could clearly reflect nothing but chance.

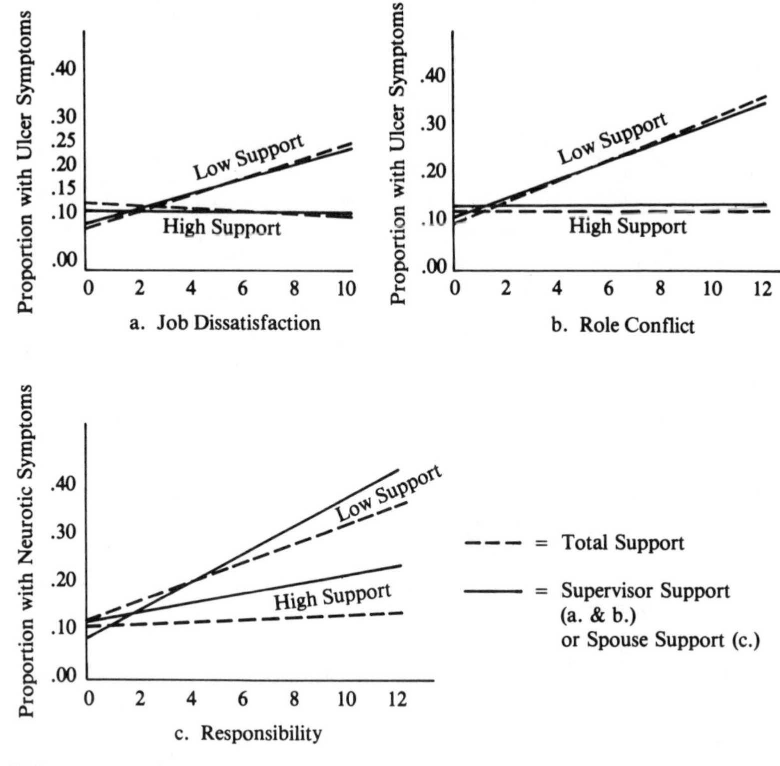

FIG. 4.2 *Buffering effects of social support on relationships between stress and health among factory workers (Based on data in House and Wells, 1978)*

Table 4.3 summarizes all of the significant buffering effects in this study, and reveals a number of important patterns. First, half (18 of 35) of the stress-health relationships examined in this study were significantly buffered by social support. Second, as already noted, the effects of the measure of total support are in almost all cases accounted for by the effects of a single salient source of support; and some sources of support, work supervisors and spouses in this study, are clearly more significant than others.

The only surprise in these results is the relative lack of effect of coworker support. Perhaps the strong effect of supervisor support and the relatively weak effect of coworker support (also evident in Table 4.2) is due to the way work is organized in this

TABLE 4.3 *Summary of significant buffering effects of social support on relationships between perceived stress and health (from House and Wells, 1978)*

	HEALTH OUTCOME				
PERCEIVED STRESS	Angina Pectoris	Ulcers	Itching and Rash	Cough and Phlegm	Neurosis
Job Satisfaction	Wife	Supervisor Total	Wife		Wife
Work Self-Esteem	Wife	Supervisor Total	Supervisor	Total	Coworker Total
Job versus Nonjob Conflict		Supervisor Friend and Relative Total			
Role Conflict		Supervisor Total	Supervisor		Supervisor Wife Total
Quality Concern					Wife
Responsi-bility				Supervisor	Wife Total
Workload		Coworker Friend and Relative Total		Supervisor	Wife

NOTE: Cell entries indicate measures of support that significantly buffer each health-stress relationship.

factory. Since many jobs are individual and machine-bound, noise levels are high, and work schedules and processes are tightly controlled by management, coworker cohesion and interaction is reduced and the potential main and buffering effects of coworkers reduced. These hunches are supported by the results of the next study considered here that show coworker support to be equally or more consequential than supervisor support in a different set of occupational settings. Finally, Table 4.3 shows that support buffers the effects of stress on ulcers and neurosis more than on other health outcomes. In fact, the results for the other health outcomes are not substantially greater than what might occur by chance. This pattern is consistent with prior research and theory that has emphasized the role of interpersonal processes in the etiology of ulcers (Susser, 1967) and neurosis (Jaco, 1970).

In sum, this study of factory workers indicates that social support can directly reduce work stress and hence indirectly improve health, while also buffering workers against the effects of stress on health. These effects, however, are produced by only one or two significant sources of support (supervisors and spouses) and in the case of the buffering effects may be limited to selected health outcomes (ulcers and neurosis) that are particularly sensitive to interpersonal processes.

STRESS, SUPPORT, AND HEALTH IN 23 OCCUPATIONS

One of the major limitations of the work by both Cobb and Kasl and House and Wells just reviewed is that the samples are limited to factory workers in one or two plants and geographic locations. The results of these studies are fairly compelling, but can they be generalized to workers in other occupations and locations? A study of support, stress, and health in 23 different occupations (Caplan et al., 1975) provides the answer, which appears to be yes. These researchers studied over 2,000 male workers in 23 occupations, from physicians to assemblers, drawn from 67 different sites or organizational affiliations in the eastern, midwestern, and the southern United States.[7] The measures of support closely parallel those in the House and Wells research, except that spouse and friend and relative support were collapsed into a single measure of home support. The measures of perceived stress in the two studies are also similar, while the health measures were comparable enough to allow some replication, as well as extension of earlier results.

The results of this study closely parallel those of the House and Wells study. Pinneau (1975, 1976) performed initial analyses of the effects of support in these data. His summary of the main

7. Most of the results reported here are based on a random subsample of about 28 workers from each of the 23 occupations (N = 636). This subsample insures that no single occupation or occupational level contributes disproportionately to the results.

effects of support on work stresses and both job-related and general psychological strains (indicative of poor mental health) corresponds closely with the results seen in Table 4.2, except that the effects of coworker support rival or exceed those of supervisor support:

> *Support from home had little effect on job stresses, while support from supervisor and from coworkers both had numerous effects on a variety of stress measures. The size of the correlations varied considerably from occupation to occupation, but the direction of significant effects was almost always as predicted. Men with high support from either supervisor or coworkers generally reported low role conflict, low role ambiguity and low future ambiguity, high participation, and good utilization of their skills.... The magnitude of these correlations were often in the .30's and sometimes in the .40's ... social support predicted significantly to low levels of psychological strain in a number of instances ... home support correlated much less often with the job dissatisfaction measures than supervisor and coworker support. Each of the* general *affective strains (i.e., depression, anxiety, and irritation) was, however, affected by both home and work support measures.*
>
> (PINNEAU, 1976, pp. 35–36)

Pinneau (1975, 1976) reported no more significant buffering effects in these data than might occur by chance. His analysis strategy for buffering effects, however, contained certain logical and technical flaws (see House and Wells, 1978; LaRocco et al., 1980). Reanalysis of these data by LaRocco et al. (1980) using methods comparable to those of House and Wells (1978) already discussed, showed that support did indeed buffer the impact of stress on general psychological strains (anxiety, depression, somatic complaints). This analysis examined the impact of nine measures of perceived work stress (for example, workload, role conflict) on three measures of general affects about the job (that is, dissatisfaction and boredom) and four general indicators of psychological strain (anxiety, depression, irritation, and somatic complaints). The impact of job-related affects on general psychological strains were also examined. These analyses revealed many more buffering effects than might occur by chance with respect to the general indicators of psychological strain, but only chance level results with respect to job related affects, most notably job dissatisfaction. In almost all cases where the impact of stress on psychological strain varied significantly across levels of support, the pattern of results indicated a pure buffering effect. Several of these effects, with respect to somatic complaints, are illustrated in Fig. 4.3. In each case, if social support from coworkers is low, somatic complaints increase as perceived stress increases; but if coworker support is high, perceived stress is not associated with increased somatic complaints. That is, *high social support can completely eliminate the deleterious impact of stress on health.*

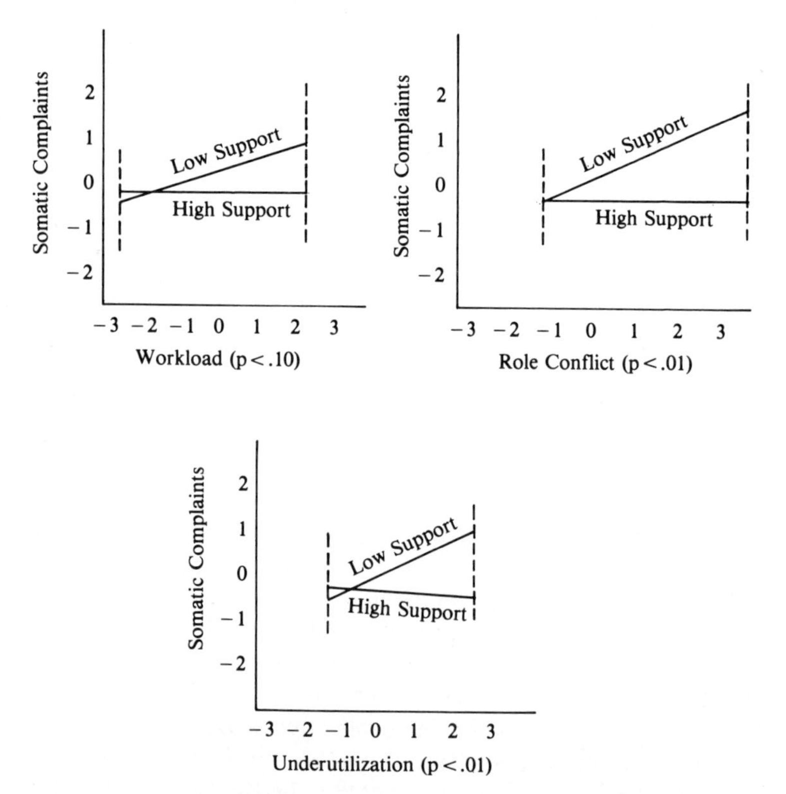

NOTE: Dotted lines delineate highest and lowest observed values of perceived stress variables.

FIG. 4.3 *Buffering effects of coworker support on relationships between perceived job stress and somatic complaints (adapted from LaRocco, House, and French, 1980, Fig. 2).*

The overall pattern of the results with respect to psychological strains is presented in Table 4.5. Twelve measures of perceived stresses and job-related affects can affect general psychological affects or strain. The first five perceived stress measures asked people directly whether they perceived aspects of the job as stressful or unpleasant. The last four were more experimental—constructed by taking the difference between people's reports of what their job was like and what they wished it to be like on four dimensions (for example, workload, role ambiguity). For a number of reasons, the last three "fit" measures did not work well in some aspects of the larger study. Nor did they produce many positive results here.

If these all four "fit" measures are excluded from consideration, there are 32 relationships in Table 4.5 of the first five perceived stresses and the three job-related affects to the four psychological strains. Of these 23 exhibit statistically significant and theoretically expected buffering effects from some form of social support. That is, 75 percent of the potential deleterious effects on psychological strain are alleviated by social support—evidence quite consistent with the Cobb and Kasl (1977) and House and Wells (1978) studies.

However, the relative importance of different sources of social support is different in Table 4.5 from the earlier studies. Of 32 relationships that might be buffered by a given source of support (excluding the "fit" variables), 10 are buffered by supervisor support, 8 by home support, and 19 by coworker support. Thus, supervisor and home support, which were the most consequential forms of support in both the Cobb and Kasl and the House and Wells studies, remain important here—buffering about one-quarter of all stress-strain relationships. But coworker support emerges in these data as about twice as important, buffering almost 60 percent (19) of the 32 possible stress-strain relationships.

Thus, these data further indicate that different sources, and perhaps types of support, are important in different contexts. In the Cobb and Kasl study of men losing their jobs, support from people outside of work, especially spouses, was critical because work-related sources of support were less available. House and Wells (1978) speculated that coworkers were not consequential sources of support in their data because the organization of work in the factory study tended to isolate workers from each other,

TABLE 4.5 *Significant buffering effects of social support on relationship of perceived stress and job-related affects to general psychological strain (compiled from data in LaRocco, House, and French, 1980)*

	Psychological Strain			
Perceived Stress/ Job-Related Affect	Depression	Irritation	Anxiety	Somatic Complaints
Perceived Stresses				
Role Conflict	Coworker Home	Supervisor Coworker Home	Coworker	Coworker Home
Future Ambiguity	Coworker	Supervisor	Supervisor	
Underutilization	Coworker			Coworker
Participation	Supervisor	Supervisor		Coworker
Workload	Coworker	Supervisor Coworker		Coworker
Workload Fit	Coworker Home	Coworker Home	Coworker Home	Coworker Home
Role Ambiguity Fit		Coworker		
Complexity Fit		Supervisor Home		
Responsibility Fit				
Job-Related Affects				
Job Dissatisfaction	Supervisor Coworker		Supervisor Coworker	
Boredom	Coworker Home		Supervisor Coworker	Supervisor Coworker Home
Workload dissatisfaction	Coworker Home	Coworker Home	Coworker	

physically and socially. Thus, supervisors became the most available and important sources of support at work. The wider range of occupations studied by LaRocco et al. (1980) included professional, managerial, supervisory, craft, and service (for example, police) workers, all of whom are often only nominally supervised and hence rely heavily on work peers and colleagues for support. Police officers get social support more from their patrol mates than from supervisors, tool and die makers from their fellow craftsmen, and physicians, scientists, and engineers from their professional colleagues (to name just a few of the occupations included in this study). In sum, all people can benefit from social support in relation to their particular occupational stresses, but who can give a person the most effective support depends on the kind of work he or she does and the kind of stresses it imposes.

One other noteworthy result of the LaRocco et al. work was the confirmation of a finding from similar studies (House, 1980; LaRocco and Jones, 1978) that social support can buffer the effects of work stress on health (as well as having main effects on health), but has primarily main effects on perceived stress and job-related affects (job dissatisfactions). Possible reasons for this will be noted next.

WORK STRESS, SOCIAL SUPPORT, AND HEALTH: SUMMARY AND IMPLICATIONS

Data from both general studies of stress, social support, and health, and more focused studies of work stress, social support, and health consistently support the proposition that social support can reduce work stress, improve health, and buffer the impact of work stress on health. Although not large, the body of data reviewed here is remarkably consistent. But not any and every form of social support will reduce every form of work stress, or enhance all aspects of health, or buffer all relationships between stress and health. Indeed, the major task for both future research and application is to specify under what conditions what kinds of social support will have what kinds of effects on stress and health. The available evidence suggests both some initial answers to this question and some promising strategies of both research and application for obtaining better answers in the future.

First, social support clearly can both reduce work stress and buffer the impact of stress on health. Since stress normally has an adverse effect on health, both of these effects constitute mechanisms through which support maintains or improves health. It is unclear, however, whether support has beneficial effects on health beyond its contribution to reducing stress and buffering the impact of stress on health. Although it seems logical that it should, the available empirical evidence does not indicate that such effects are strong. For example, none of the three major studies of work stress, social support, and health find much evidence for such effects.

Second, it appears, at least in occupational settings, that the stress-reducing effects of social support are quite general (see Table 4.2 and Pinneau, pp. 35–36) while the buffering effects of support are more limited. In the Cobb and Kasl study, support buffered a wide range of work-role deprivations and mental health outcomes, but had only limited effects on physiological variables and medical conditions. In the House and Wells research, support primarily buffered the effects of perceived stress on ulcer and neurotic symptoms. And in the LaRocco et al reanalysis of Pinneau's data, mental health symptoms were again the most affected. There are a number of possible reasons for this selectivity, although at present there is no basis for choosing among them. These health outcomes may be more susceptible than others to the effects of interpersonal relationships, especially when people are under stress. These kinds of outcomes may also be the type that are relatively manifest and hence elicit support. That is, support has generally been conceived and measured as a resource that can be drawn in times of need. But potential supporters may give support primarily in response to manifest evidence of strain and symptomatology. If support is, in the words of John R. P. French Jr. (personal communication), "strain-responsive," it will most effectively buffer people against strains that are both manifest to their potential supporters and considered legitimate grounds for support.

Third, note that most of the studies showing effects of support have measured either (1) the mere presence or absence of relationships (for example, Berkman and Syme) or (2) the emotional supportiveness of relationships. Thus it appears that the presence of significant, emotionally supportive relationships is responsible for most of the documented effects of social support. How emotional support relates to the other forms of support discussed in Chapter 2 and whether these other forms of support have similar effects are issues for further empirical research.

The existing research indicates that "the crucial quantitative difference in the number of supportive relationships is between zero and one—between those who must face stressful events with no close relationship and those who are supported by at least one such relationship" (Kahn and Antonucci, 1980). That is, support from *one* significant other can be quite effective in mitigating the effects of stress on health, and support from additional sources may have little or no additional benefits. The House and Wells data clearly show that a single source of support is responsible for almost all buffering effects, even those produced by a composite measure of total social support. Studies that use composite measures of support (e.g., Berkman and Syme, 1979; Cobb and Kasl, 1977) generally give primary weight in their measure to one or two significant relationships (for example, spouse, close friend). The efficacy of a single confidant or supporter has also been documented in a range of other research (Brown, Bhrolehain, and Harris, 1975; Lowenthal and Haven, 1968).

However, which signicant other provides the support that reduces stress, improves health, or buffers the impact of stress on health will vary across individuals and situations. Work-related sources of support (work supervisors and coworkers) are most effective in both reducing occupational stress and buffering the impact of such stress on health, although support from spouses is also important in buffering the impact of work stress, especially on general affective and psychological states. In some occupational settings supervisors are likely to be the most effective sources of support, while coworkers are most effective in others. Thus, researchers and practitioners must specify the relevant stresses and health outcomes to know which sources of support are likely to be most effective.

In sum, since the evidence indicates that social support can reduce stress, buffer the impact of stress on health, and improve health (either as a result of the first two effects or directly), it is time to begin thinking about experimental applications of this knowledge. And this current knowledge suggests some clear directions for such applications. What researchers have so far, however, is mainly regarding the effects of *perceived* emotional support. It is clear that many different types of workers in many different situations are better off if they perceive others as willing and able to help (especially in an emotional sense) with work-related problems. But how do people come to have that perception? That is, what are the social, interpersonal, and personal factors that promote or inhibit the development of a subjective perception of social support relevant to work stress. This much neglected issue in the study of social support must be addressed if effective applied programs for enhancing social support are to be developed.

III

USING SOCIAL SUPPORT TO REDUCE STRESS AND IMPROVE HEALTH

5

THE DETERMINANTS OF SOCIAL SUPPORT

The focus of this and the concluding chapter moves increasingly from issues of research to issues of practical application, still recognizing the necessary and natural links between theory research and practice noted in the Introduction. Understanding how social support, as it naturally occurs in people's lives, operates to reduce work stress and improve health is in many ways analogous to understanding how people's natural immunity systems of their bodies protect them against the ravages of disease (Cassel, 1976). An important practical benefit of understanding such natural immunity systems is learning how to enhance, and even synthesize, the essential health-protective elements in such systems.

It is a big step, however, from understanding how natural immunity works to intentionally and artificially synthesizing or enhancing the elements of such immunity. For example, in recent years biomedical scientists have succeeded in isolating a natural substance known as *interferon* and observing its ability to destroy malignant cells in the body while leaving normal cells unharmed. Thus, interferon has become a major hope in the search for more effective chemotherapy against cancer. The principal stumbling block to widespread use of interferon in clinical trials and, if the trials are successful, in standard clinical practice is that scientists can at this point neither stimulate natural bodily production of interferon nor synthesize or create it artificially. Thus, even if current clinical trials of interferon proved it could eradicate cancer, physicians could not make widespread practical use of this without greater scientific knowledge about how to produce this substance artificially or to enhance the natural production of it in the bodies of people with cancer.

Similarly, if we are to make widespread use of social support to reduce work stress and improve health, we must know how to synthesize, create, or enhance such support where it is weak or nonexistent. This knowledge requires understanding the causes or determinants of social support, and especially identifying those causes or determinants of support that can serve as points of intervention or synthesis. Although researchers probably know more about the factors that produce social support than about the factors that produce interferon, in research on social support they have been preoccupied with the nature of social support and its effects on stress and health. This preoccupation is especially true of work on occupational stress and social support. Thus, understanding of the causes and determinants of support, and hence of how to enhance support, is not as great as it could or should be.[1] Nevertheless, enough is known both to begin to make practical use of this knowledge and to identify fruitful avenues for further research. This chapter will first identify the kinds of factors that logically must determine the level of social support people experience at a given time and place, and then consider what is known concretely about the importance and impact of various specific causative factors. This discussion leads naturally into suggestions for practical action, which are alluded to in this chapter and developed more fully in Chapter 6.

BASIC PROCESSES AND FACTORS GENERATING SOCIAL SUPPORT

To understand what makes a person, Pat, feel that another, Oscar, is a source of one or more types of social support, researchers must

---

1. A major exception to this generalization is the work of Gerald Caplan and his colleagues, which has developed with the framework of community health and been oriented to ways of providing and enhancing social support, especially from natural social networks. Yet even this work has been limited in the range of causal factors it has considered and the types of stressful situations it has addressed.

understand an interpersonal process like that depicted in Fig. 5.1. This model suggests that some of what Oscar does and feels (Box 1 of Fig. 5.1) is directly perceived by Pat (Box 2). For example, Oscar says, "You're one of my best friends," and Pat may hear this, not hear it, or mishear it (e.g., Pat thinks Oscar said, "You *were* one of my best friends"). What Pat perceives is a function not only of Oscar's behavior but also of characteristics of Oscar, Pat, or their relationship and of the social and cultural context in which their interaction occurs. These factors may directly affect Oscar's behavior or Pat's perception or may condition the relation between the behavior and its perception. For example, Pat may be more likely to mishear what Oscar said if Pat is hard of hearing or Oscar is soft-spoken or they are in a noisy place or they speak different languages or just met for the first time.

However Pat perceives Oscar and what Oscar has said or done, Pat must then interpret, or give meaning to, Oscar and Oscar's behavior (Box 3 of Fig. 5.1). This part of the process is critical because even if Pat accurately perceives what Oscar is and what Oscar does, the interpretation of Oscar and Oscar's acts may vary radically depending upon the kind of social, cultural, and personal factors just noted. Just because Oscar says to Pat, "You're one of my best friends," does not mean that Pat will take this statement at face value or interpret it as indicative of social support. People and behaviors must be interpreted in the context of both the immediate (or microscopic) situation and the larger (or macroscopic) sociocultural milieu. If Pat sees Oscar as attempting to ingratiate or manipulate her (because, for example, Oscar and Pat are members of opposing groups or Oscar badly wants something that Pat has), Oscar's statement may be interpreted as unfriendly and nonsupportive. How supported Pat feels will also depend on how much support she wants or needs. On the basis of this interpretation Pat will exhibit behaviors or affects (for example, Pat looks happy) that Oscar may perceive, misperceive, or fail to perceive, and which Oscar in turn will interpret and respond to with behavior or affect (Boxes 4, 5, and 6 of Fig. 5.1). Thus, giving and receiving social support involves a process of mutual social interaction; and the major variables in this process and the relations between these variables are affected by a range of social, cultural, and personal factors.

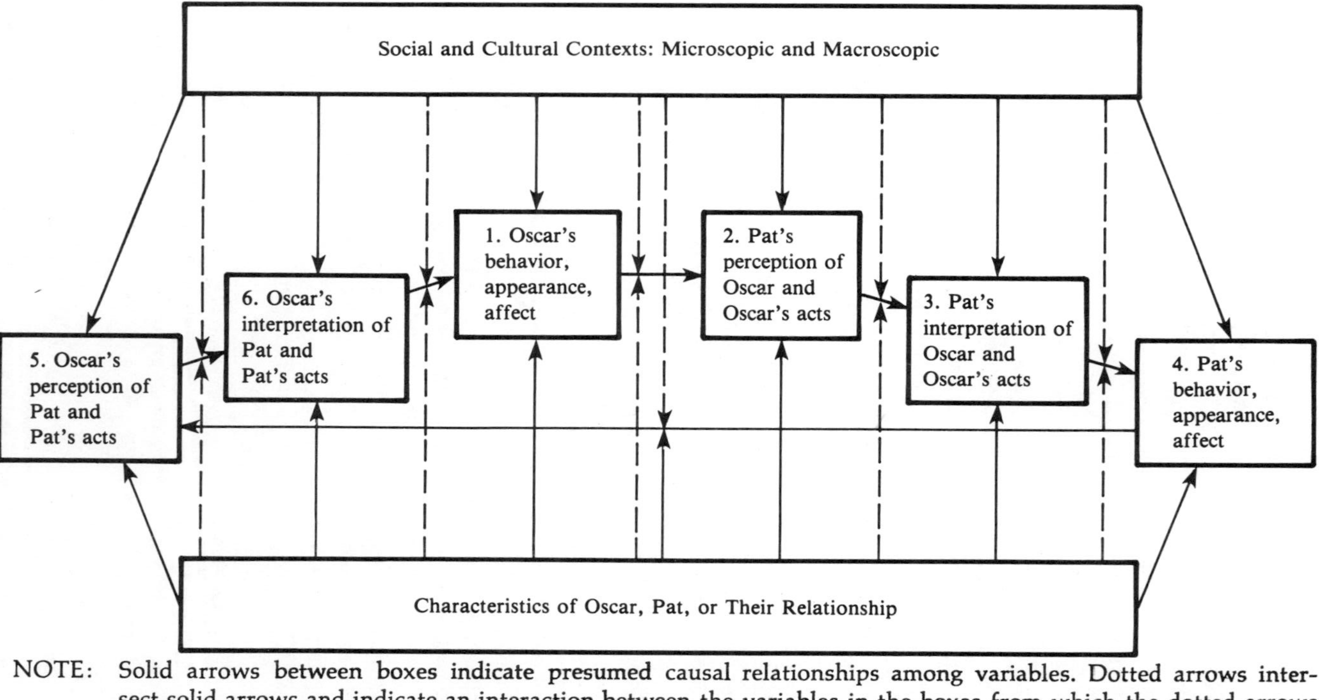

NOTE: Solid arrows between boxes indicate presumed causal relationships among variables. Dotted arrows intersect solid arrows and indicate an interaction between the variables in the boxes from which the dotted arrows emanate and the variables in the box at the beginning of the solid arrow in predicting variables in the box at the head of the solid arrow.

FIG. 5.1 *Hypothetical model of interaction process between Oscar and Pat that generates feelings of social support*

Researchers are just beginning to understand this potentially complex process. For many purposes it may turn out to be less complex than it potentially could be. We must begin by considering those ideas and facts that give us the greatest insight into the process at this point, while noting priorities for future inquiry. Three major foci for any effort to understand, and hence alter, levels of support are:

- *characteristics of individuals,* which facilitate or impede their ability to give or receive social support;
- *properties of relationships,* which may facilitate or inhibit the giving or receiving of social support;
- *social or cultural conditions,* which foster or discourage the giving or receiving of social support (these effects will sometimes operate via the effects of such conditions on characteristics of persons or the nature of interpersonal relationships).

All three sets of factors and their interrelations will be considered, with a highlight on the importnce of the third set of conditions, since they are often overlooked in current discussions of the determinants of social support and strategies for enhancing social support.

Throughout the discussion, remember that supportive relationships, especially naturally occurring ones, are likely to be reciprocal exchange relationships in which, for example, Oscar supports Pat and Pat supports Oscar. We can demand or buy support in a unilateral way from professionals and specialists whose job it is to provide certain kinds of support (for example, service or care givers, clergymen, mental health workers). But informal sources of support, those with the most potential for preventing work stress and its deleterious consequences, are likely to be involved in mutual exchange relationships (for example, between friends, spouses, coworkers.)[2] The currency of such exchanges need not always or

2. Work supervisors, who will be a focal point of discussion in this chapter and the next, will probably be somewhere between the more formal and informal sources noted here. Providing support should be part of the supervisor's job, but the nature and extent of supervisor support will undoubtedly depend on a process of mutual exchange between supervisors and subordinates. Mutual support between supervisors and subordinates will undoubtedly be most effective for both, and will evolve most easily in the context of a mutually respectful, relatively equalitarian, relationship.

equally be support on both sides, but generally the parties in such relationships will each both give and receive support. Conceiving of support as a mutual, exchange relationship reminds us that we can change the supportiveness of Oscar toward Pat not only by influencing Oscar, but also by influencing Pat, and especially by influencing the quality of their relationship.

Finally, thus far support has been discussed in simple terms, that is, a relationship between Pat and Oscar. But each Pat may have many "Oscars" who actually or potentially provide social support. Social network analysts have emphasized the importance of the structure of relationships between the "Oscars," as well as their relationships with Pat, for understanding the nature and determinants of social support; these issues will be considered here. It is also important to distinguish between different types of Oscars (as is done, for example, in the classification of sources of support in Table 2.2). For the purposes of this chapter and the next, the most basic distinction is that between work-related and nonwork-related sources of support, since these have different effects (see Chapters 3 and 4) and different causes. For a variety of reasons the discussion will focus initially on work-related sources of support, turning then to nonwork-related sources (especially spouses). Both formal or professional and informal sources are considered.

WORK-RELATED SOURCES OF SUPPORT

Evidence reviewed in Chapter 4 showed work-related sources of support to be most effective in reducing work stress and buffering the impact of work stress on social support. Thus, work-related sources of support receive primary consideration in the discussion of applied intervention and research programs in Chapter 6. Hence they should also receive priority in our efforts to understand the determinants of social support.

Supervisor support

Work supervisors are a consequential source of support in research reviewed in Chapter 4, especially in situations where the opportunities for cohesive interaction with coworkers may be limited. Limited interaction with coworkers is a common feature of many industrial and office jobs and also of jobs where the worker operates as a mobile, semiautonomous unit (many types of sales and service jobs). Thus, supervisors are identified in Chapter 6 as a primary target for efforts to enhance work-related social support.

But what factors determine the supportiveness of supervisors? Little direct evidence is available about such factors. Research on work stress and social support has looked at the effects of supervisors who are perceived as supportive on both work stress and health, but has not explicitly asked two important and distinct questions: (1) *what is it that supportive supervisors do* that makes their subordinates perceive them as supportive? and (2) *what causes them to act in a supportive manner?* Some answers to these questions are, however, implicit in some data from existing studies of support and from the more general literature on organizational behavior.

Social support has been identified as a major component of effective supervision in a number of organizational theories, most notably the human relations approach of Likert (1961) and others. Likert's research indicates what supportive supervisors do that leads them to be perceived as supportive. He reports that "supervisors who have the most favorable and cooperative attitudes in their work groups" are perceived by their subordinates as follows:

> *He is* supportive, *friendly, and helpful rather than hostile . . . genuinely interested in the well-being of subordinates. . . . He sees that each subordinate is well-trained for his particular job. He endeavors to help subordinates be promoted . . . giving them relevant experience and coaching whenever the opportunity offers. . . . He coaches and assists employees whose performance is below standard. In the case of a subordinate who is clearly misplaced and unable to do his job satisfactorily, he endeavors to find a position well suited to that employee's abilities and arranges to have the employee transferred to it.*
>
> (LIKERT, 1961, p. 101)

In sum, the good "human relations" supervisor provides what I have called emotional, informational, instrumental, and appraisal support. The specific supervisory behaviors Likert describes, like those enumerated by Gottlieb in Chapter 2 (see Table 2.1), provide tangible illustration of these abstract concepts. Indeed Likert (1961, pp. 102–103) refers to the integrating principle of his theory as "the principle of supportive relationships."

Likert and other human relations writers are less clear on

exactly how supervisors get this way, much as students of psychotherapy have some agreement that good therapists are warm and empathic but are unclear how they get that way. Presumably some people are more likely to be supportive than others as a result of their prior social and psychological experience and development. If we could validly assess the traits giving rise to a supportive personal orientation, we could select supervisors and therapists at least in part on the basis of these traits. We could also try to provide more training and experience in giving and receiving social support, as well as other interpersonal skills, in the socialization and education of children in our society. As will be noted below, United States society, and perhaps western, industrial society as a whole, may be relatively deficient in this regard.

Many mature adults can be trained in some of these skills, as the literature on organizational change and psychotherapy shows (Katz and Kahn, 1978, Chapter 19; Durlak, 1979). In fact, Durlak (1979) concludes from a review of the literature on counseling that, with appropriate training and supervision, paraprofessional or lay counselors can provide as effective counseling in a number of situations as mental health professionals (psychiatrists, psychologists, and social workers). A number of field experiments have been done in which organizations were changed in the direction of Likert's human relations model, and a component of each of these has been changing the supportiveness of supervisors (see Katz and Kahn, 1978, Chapter 19). Thus, some key skills for giving support apparently can be acquired through training.

However, their effective and sustained application in supervisory-subordinate relationships is not as simple as it may first appear. Training people to be supportive involves much more than a simple injunction to "be supportive or nice." In fact, naive efforts to follow that injunction can often do more harm than good. A growing body of social psychological research suggests that well-intended efforts to help or sympathize with people in distress sometimes leave the distressed persons worse off rather than better off. Coates and Wortman (1979), for example, review a number of studies of how people respond to others in substantial psychological or physical distress (for example, depressed individuals, cancer patients). The most common responses, intended to help the distressed individuals, are either to avoid saying things that might increase distress or to say things intended to reassure or cheer up the distressed persons. "Unfortunately, available evidence suggests that these initial attempts to control (a distressed) person's feelings will probably not have the desired effect" (Coates and Wortman, 1979, p. 153). The distressed people feel no better after being "helped," and often feel worse.

The reasons for this boomerang effect are complex and not fully understood. Sometimes such efforts appear to be perceived as overly cheery and hence insensitive or unrealistic. Other times, the recipients may find the concern and attention of others quite rewarding and only wallow further in their distress. Thus the effects of naive helpfulness are often bad for the recipient. Furthermore, they soon leave the helper embittered as well. When an individual fails to respond to a helper's aid, sooner or later the helper responds negatively toward that individual (covertly, if not overtly).

These negative outcomes can be avoided, provided that support givers are properly trained. Empathic listening and understanding and real concern for the other's welfare are important components of support often absent in naive efforts to "make people feel better." Thus, although the skills to be taught in any effort to increase supportive behavior may not prove terribly difficult to communicate or acquire, they are real skills that can probably best be acquired with professional help.[3]

Even more problematic than supervisory acquisition of supportive behavior is how to sustain it. What supervisors do depends not only on what they know how to do and their personal values or desires, but also on the larger interpersonal, social, and organizational context in which they work. Katz and Kahn (1978, Chapter 19) emphasize that efforts to change individuals' work-related attitudes or behavior are likely to have limited or short-lived success unless their social environment is changed to reinforce the newly acquired attitudes and behavior. In the case of managers

---

3. Coates and Wortman (1979) suggest that the likelihood of a boomerang effect increases with the extent to which the helper wants to *control* the behavior and affects of the recipients of help, rather than helping the others regain control of their own affairs. Since the business of managers and supervisors is control, there may be a special need for training here. As Levi (1978) notes, organization expends surprisingly little effort to train people in how to supervise, yet the impact of supervision is great (Berg, Freedman, and Freedman, 1978).

and supervisors, the most important determinants of their behavior appear to be what their organizations, especially the higher levels of management, advocate, value, and actually reward.

Berg et al. (1978, Chapter 11) paint a rather gloomy picture of the long-term success of experimental efforts to increase workers' participation in organizations and the quality of working life. Although dramatic short-term changes and improvements have sometimes been achieved, there is little evidence that these changes have persisted for more than a few years, at least in cases where changes in managerial behavior (as opposed, say, to the technology of work) were the primary source of change. Most experiments are undertaken in one or a few units of a larger organization, and broader organizational and managerial support for the experimental changes often wanes over time in the face of pressures for improved organizational or individual productivity and performance. Berg et al. (1978, pp. 162–164) review a case of work reform in which the

> *plant manager's initial support of participative management . . . shifted toward a philosophy of "manage in whatever way is comfortable for you as long as you get good results." Since many managers view their present position as just a stepping stone to a better job, they have misgivings about the usefulness of reforms, about the implications of reforms for their own status, and about long-term commitments.*
>
> (BERG ET AL., 1978, p. 164)

Thus, sustained changes in supervisory or managerial behavior, including increased emphasis on social support, are likely to occur only in the context of a broad organizational participation in, and support for, such change.

The ability of supervisors to provide social support is likely also to depend on the number of employees they supervise, the nature of the supervisory task, and the relation among the supervised employees. The work of Likert and others implies that supportive supervision can be instituted most easily where the number of subordinates, and the nature of their tasks, is such as to allow or even require participation by groups of subordinates in planning and organizing work activities.

Coworker support

More concrete evidence regarding the conditions that promote or inhibit the develop of coworker support is available. One of the key attributes of Likert's ideal organizational system is that each person be a member of one or more supportive, effective work groups. The formation of such groups is perhaps the major goal of supportive, participative supervision. Thus, coworker support should vary directly with the degree to which supervisors model supportive behavior and use participative, group-oriented methods of supervision. Likert (1961) and others present evidence confirming this proposition.

Empirical evidence also suggests, however, that the structure of the organization and the jobs within it also have a strong influence on coworker support. The weak effect of coworker support on stress and health in the House and Wells study of factory workers (see Chapter 4) was attributed to a highly individuated structure of work in that factory. That study also provides more direct evidence on this issue. Workers in jobs paid on an individual piecework basis (jobs that are necessarily highly separate from others) reported lower levels of coworker support than workers in jobs paid on an hourly basis (which involve more interaction and cooperation with others). The pay system, by fostering production competition among workers, may have adverse effects over and above the effects of highly separated jobs. Caplan et al. (1975, p. 187) show that factory workers who work quite independently of others (for example, assembly-line workers and machine tenders) report levels of coworker support (and to a lesser degree supervisor support) much below the average for all workers. In contrast, operators of continuous process equipment who have more freedom to interact with coworkers and supervisors report above average levels of support.

A final factor in levels of coworker (and perhaps also supervisor) support is the basic nature of employee–organization relationship. The high rates of chronic disease and low levels of social support are not necessarily concomitants of an industrial economy. A number of observers have been struck by the low coronary heart disease rate in Japan and by the fact that Japanese who migrate to the United States lose their relative immunity to heart disease in proportion to the degree that they give up traditional

Japanese social and cultural ties (Matsumoto, 1970; Marmot, 1975). They have speculated that these differences stem from a much higher level of social support, which characterizes Japanese society and industry:

> *The etiology of coronary heart disease is multiple and complex, but in urban-industrial Japan, the in-group work community of the individual, with its institutional stress-reducing strategies, plays an important role in decreasing the frequency of the disease. . . . In spite of rapid social change . . . observers agree that Japan has not moved from group values toward individualism but rather retains strong emphasis on collectivity orientations with the in-group. . . . The immediate work group is of primary importance. . . . Group membership is continuous over the years. Both the individual and the firm look upon his employment as a lifetime commitment. . . . Among his fellow employees, the Japanese individual can relax, argue, criticize and be obstinate without endangering relations. . . . In both labor and leisure the Japanese individual is involved primarily with his fellow workers. . . . with built-in social techniques and maneuvers for diminishing tension.*
>
> (MATSUMOTO, 1970, pp. 14, 15, 16, 18, 19, 21)

Matsumoto reviews specific characteristics of Japanese social patterns and practices in both work and leisure that support his speculations, and notes the need for focused research on these issues (along with other differences between Japan and America, such as diet, which may account for the better health of the Japanese). Our society clearly has something to learn from other societies about how to enhance social support at work and more generally.

In sum, levels of coworker support seem to reflect the influence of supervisory behavior, organizational and job structure, and the values and structure of the organization and the larger society. In this regard, the competitive, individualistic ethos of American society, and especially the American economy, appears inimical to the development of strong social support networks. Intensely competitive, evaluative, or unequal social relationships are not conducive to supportive relationships. Nor, are they necessarily productive of individual or organizational achievement[4] (see Chapter 6).

The potential role of unions

Given the hazards of competition between workers and the divergence in many instances between the interests of subordinates and superiors, unions provide another potential focus for efforts to enhance social support, though relevant, of course, only in unionized work settings. The members of the union have a normally cooperative relationship with each other, and their stewards and officials are more exclusively concerned with the welfare of workers than are supervisors or employers. Union stewards, then, may replace or complement supervisors as sources of social support. Unions also constitute an alternative to employers as the organizational mechanism for promoting social support among peers, colleagues, and coworkers. As an organization the union can also promote and support efforts to change managerial practices and organizational structures in directions that facilitate support. As with employers, the unions, to enhance support, require not only that appropriate people acquire appropriate training and skills, but also that the union as a whole (that is, at all levels) have a clear commitment in this direction.

Formal sources of work-related support

The factors influencing levels of informal work-related social support have been emphasized because supervisors and coworkers are likely to be the most readily available sources of support for individual workers. However, it is important to recognize that social support can be, should be, and is provided in more formal ways by employers, unions, and governmental agencies concerned with workers and organizations. In another volume in this series,

---

4. Social psychologists have long recognized, however, that competition *between groups* can increase cohesion or support within those groups, although it may heighten conflict and animosity between groups. Better still, rather than having groups compete with each other they can cooperate in trying to achieve a common or superordinate goal (Sherif and Sherif, 1953). In any case, intense individualistic competition is clearly destructive of supportive relationships among the competitors. Cooperation within and between groups of workers will facilitate the giving or receiving of support.

Warshaw (1979) reviews a large number of such programs for managing stress. For the most part, such programs have tended to be therapeutic, rather than preventive in nature. That is, companies and unions have become increasingly interested in developing mechanisms for identifying work-stress-related physical, psychological, and behavioral disorders (alcohol and drug abuse, absenteeism, heart disease, anxiety, and depression) and providing treatment or referring employees to appropriate sources of treatment for their problems. Much of the treatment provided involves provision of information, appraisal, and instrumental or emotional support by health service workers.

Two common, but not inherent, limitations of these programs is their therapeutic, rather than preventive, nature, and their focus on the individual rather than organization as the locus for preventing and/or treating work stress and its deleterious health consequences. These programs can be made more preventive by reaching out to workers with high potential for experiencing work stress. For example, workers experiencing major occupational change (due either to their own mobility or change in the organization around them) are one such group. Organizational or job groups that have manifested high levels of stress-related disorders in the past are another potential target. In such cases preventive supportive services can be provided to individuals in the form of education, group discussion, or individual counseling. Even more desirable would be attempts to modify aspects of the situation or organization that are producing stress (for example, modification of organizational policies and requirements to make them less stressful for individuals). Efforts could also be made to strengthen the natural support systems of workers in high risk groups, or more generally. Warshaw (1979) indicates how some individually oriented therapeutic programs for dealing with work stress can be modified to embody a more preventive and organizational approach.

No matter how effective preventive efforts are, however, the need will continue for more formal sources of all types of social support available to employees who require such support to reduce their levels of work stress and/or compensate for or buffer the impact of stress on health. The more effectively such efforts are integrated with informal support systems, the more efficient and effective they are likely to be. Better informal support systems at work could deal with many of the less serious problems now confronting formal sources of help, allowing the generally more highly trained formal helpers to concentrate on dealing with more severe cases, and to carry out overall planning evaluation activities with respect to the alleviation of work stress and its effects on health.

Conclusion

Although a great deal is not known about what makes some work environments more supportive than others, researchers have identified a set of factors that should be the focus of both more intensive research and initial efforts at application. The supportiveness of supervisors, coworkers, or other informal sources of support at work is a function of both the characteristics that individuals bring with them to their work role and those that are acquired and reinforced in the work setting. It is clear that the United States society could do more to foster in all individuals from an early age the capacity to give and receive social support. However, a number of attributes of work organizations appear to affect the levels of support available. The structure of the organization and of the jobs in it can enhance or inhibit the potential for support. Some empirical evidence indicates that specialization and fractionation of jobs and the creation of highly isolated work roles is deleterious to levels of both supervisor and coworker support. Conversely, participation in work that is cooperatively interdependent work, rather than competitive or highly independent, appears to foster social support. Finally, it is clearly possible to train supervisors, union stewards, and employees themselves in giving and receiving social support. The extent to which the organizational, technological, and personnel policies of an organization tend to foster or dampen levels of support appears to be largely a function of the goals and priorities of the organization's top management and the larger economic system.

NONWORK SOURCES OF SUPPORT

A good deal of the evidence in Chapter 4 indicates that persons outside of work, especially spouses and perhaps close friends or relatives, can be effective in buffering the impact of work stress on mental and physical health. Such support may also have some directly beneficial health consequences, although it generally will have little effect on levels of work stress. For reasons discussed in the remainder of this chapter and in Chapter 6, I feel that nonwork sources of stress should not be expected to play the major role in buffering or compensating for the effects of work-related stress on health. However, it is important to recognize the contribution that nonwork sources can make in this area, and hence to be aware of the factors, including work-related factors, that determine the level of nonwork sources of support available to workers.

Although recent work has focused on social support as an independent variable affecting stress and health, Kaplan et al. (1977, p. 47) point out that early students of support, most notably Durkheim, were equally or more interested in "social support as a dependent variable, that is; what types of conditions affect or predict the level of social integration" of a society or a community? The availability of nonwork-related sources of support will be a function of the general level of social integration, and more particularly the integration and supportiveness of nuclear families (especially spouses) and networks of friends and relatives. This social integration will, in turn, be a function of (a) broad social and community characteristics, (b) the nature of particular families, marriages, and social networks, and (c) the work relationships and involvements of these persons and groups. The following discussion considers each of these sets of causes or determinants. No systematic empirical study of the macrosocial and microsocial determinants of support has been made, but relevant evidence can be pulled together from a variety of sources. In the midst of this process is a major paradox—levels of nonwork-related social support are adversely affected by the very occupational stresses that such support is expected to buffer people against.[5]

Social and community contexts

One of the truisms of modern times asserts that population growth, industrialization, and bureaucratization have severely weakened the strength of intimate social ties. The combination of increased social and geographic mobility and the growing size of the

---

5. This paradox is equally applicable to work-related sources of support. For example, plant closings, reductions in force, overload crises, managerial successions, and transfers disrupt the very social support systems in the workplace that could most help people adjust to these stressful situations.

communities and organizations in which people spend much of their lives has, some say, made it difficult to develop and maintain close relationships with family and friends. In the words of Vance Packard (1972), the United States has become a "nation of strangers." One aspect of this presumed massive decline in social integration would be a decline in levels of social support available to people from family, friends, and kin. Although direct data on social support is sparse, a wide range of systematic social science data suggest that the picture that Packard and others paint of social disintegration is grossly overdrawn. Hawley (1973) and Fischer (1973) suggest that a balanced reading of available evidence indicates that the rate of social change and mobility has not accelerated as drastically as Packard and others believe, while intimate social relations have survived very well in the midst of large cities and organizations.

Still, clear trends such as the undeniable rise in the divorce rate and the waning of rural and small town life suggest that all is not totally well. In their study of plant closings discussed in Chapter 3, Cobb and Kasl came to be interested in social support because of differences they observed in the effects the closing of a plant in a rural setting compared to one in an urban one:

> *A separate analysis of the social context of the two companies (Gore, 1973) has revealed that in the urban setting, where men lived scattered throughout the city, the plant itself was an important focus of a sense of community and social support. With the plant closing, the "community" died (Slote, 1969). But in the rural setting, the small town itself and the people in it were the major source of a sense of community and social support for the men. When the plant closed down, the community and its social organization remained largely intact, and social interaction with former co-workers who were friends were not so severely disrupted.*
>
> (COBB AND KASL, 1979, p. 174)

Blauner (1964) noted the importance of the social integration of small company towns in compensating somewhat for the alienating effects of work in the textile industry in the South. However, Cobb and Kasl (1977) found that unemployment was more prolonged in the rural community, and Fischer (1976) shows that larger communities and organizations can provide access to a much wider range of potential groups and resources for individuals with special problems and characteristics. Mutual self-help organizations, for example, are likely to be more prevalent in large communities and provide an important source of social support for people with many work-related problems such as alcoholism or drug abuse (see Caplan and Killilea, 1976). In sum, the size, social structure, and cultural climate of communities can affect the level of social support available to workers from nonworker sources, but the same variable (community size) may have quite different effects on different sources and types of support. This issue merits careful attention in future research and experiments in application.

The nature of marriage, families, and friends

Different kinds of families and friendship networks are likely to be differentially willing or able to provide support relevant to work-related problems. For years both laymen and social scientists assumed that a major function of nuclear families, and especially wives, was to provide social support that would buffer the husband and father against the stresses and strains of his work role. In many occupations, the wife and family were seen as partners in the occupational endeavor, as epitomized in the corporation wives described by William H. Whyte (1956). How supportive and helpful wives and family actually were was undoubtedly highly variable, and the severe strains of the traditional marital roles and expectations on both men and women have been clearly recognized in later social science, social commentary, and fiction (for example, Friedan, 1963).

The phenomenon of the corporation wife is still prevalent (Kanter, 1977a). The rapid increases in single adult households and the increased proportion of women working even among married women has, however, altered the potential for the family to fulfill supportive functions vis à vis the world of work. The dual income and dual career marriage clearly differs from the traditional single wage-earner marriage in the types of social support that it can give and that it needs to receive. The support of women by men is as necessary as the support of men by women. Research on work stress, especially in relation to social support, has had a sexist bias in that the subjects are almost always males and spouse support almost always means wife support. In some ways more equalitarian marriages may be better sources of work-related social support, in others, worse. Where both members of the couple work, their ability to share and empathize with the other's work experience may be greater, but two jobs coupled with marital and family responsibilities may leave little time or energy for such sharing and empathy. Again, clear data are lacking, but the impact of the changing nature of the nuclear family on social support must be borne in mind in future research and application.

The nature of friendship and kinship relationships and their relation to the nuclear family are also highly variable in ways relevant to the potential of nonwork sources of support to buffer work-related stresses. As Cobb and Kasl found, the degree to which coworkers are also friends outside of work will affect the level and type of support both may provide. Sharing of similar, if not common, work experiences should increase the ability of friends and relatives to provide all types of social support for work-related problems. Sharing of interests and experiences may not be enough. To give and receive effective social support vis à vis work-related stresses, friends and relatives must be able to communicate regarding these stresses, but they must also be able to communicate in a supportive and nondefensive manner. Competition and concern over relative occupational and social status implicitly or explicitly characterizes many relationships with friends and relatives. To admit occupational unhappiness or weakness may be very difficult in such quarters, and hence potential sources of support are unavailable. In other cases, the distinction between work and nonwork phases of a person's life may be so sharp as to preclude carrying one over into the other. Again greater attention must be given to such issues in future work on social support.

The impact of work on nonwork support

One of the major determinants of the nature and availability of nonwork support is the nature of work itself. Jobs are people's major sources of income and prestige in the eyes of the community and often constrain where people live. Thus, as noted at several points, different occupations and industries are conducive to different levels and types of nonwork relationships and social support. Whether a family has one, two, or more wage earners markedly affects family relationships. Certain occupations, industries, or plants may promote supportive nonwork relationships and make them more (or less) relevant to work stresses. Workers in single-company or single-industry communities are likely to have highly overlapping work and nonwork networks, and friends and relatives who are not also coworkers may still be highly knowledgable about, and attuned to, work-related problems. Workers who work unusual schedules (such as policemen, typographers, or night shift workers) are more likely to be friendly with each other outside of work as well as at work (see Lipset, Trow, and Coleman, 1956), though their work schedules may disrupt their relations with family and other friends.

More importantly, however, I think it is important to recognize a difficult paradox—work stress often adversely affects the very network of relationships that could buffer individuals against the deleterious effects of such stress. Kanter (1977b, pp. 31–51), in an excellent overview of the reciprocal relationship between work and family in the United States, notes a variety of evidence that occupational conditions often considered stressful tend to disrupt and weaken relationships between workers and their families, thus making the family a less potent source of social support than it might otherwise have been. The family life of workers who feel unhappy or tense on the job are more likely to be characterized by unhappiness, tension, and hostility than the family of workers who were less stressed and more satisfied at work.

The amount and scheduling of working time can also strain family relationships. Excessive time devoted to work must detract from time available for family relationships, and this can have debilitating effects on the emotional climate of family life. Comprehensive, systematic data are lacking, but one only has to read the newspaper reports from Washington to find evidence of the deleterious effects of excessive occupational involvement on marital and family life. Divorces, or other signs of family strain, are common among high-level government officials who work long hours for five, six, or even seven day a week.

Not only the amount of work but also the timing of work is a problem. Mott, Mann, McLaughlin, and Warwick (1965) documented the problems male workers on night and evening shifts had fulfilling the obligations of their roles as husbands and fathers, while House (1980) found that shiftwork was associated with lower perceived emotional support from spouses.

In sum, the very occupational pressures and tensions that require social support to alleviate them or buffer their effects may diminish the potential for obtaining such support from family. Similar effects probably occur with respect to relationships with friends and relatives as well. Thus, work stress appears to be one important determinant of the level of nonwork social support available.

Nonwork sources of professional support

Obviously, nonwork sources of professional support are a major resource for many persons confronting occupational stress. Even more than similar support systems at work, however, these supports are likely to operate in response to deleterious physical, psychological, or behavioral outcomes of stress rather than to prevent such outcomes. It is possible, of course, for such professional sources of support to take a more preventive approach to work stress, for example, a community mental health center consulting with work organizations. Because of their largely therapeutic nature and somewhat lesser relevance to work stress, however, nonwork sources of professional support will be discussed no further here. There is a large literature on this issue in the area of community mental health and the delivery of health services.

SUMMARY AND CONCLUSION

Although researchers have discovered much in recent years about the impact of social support on stress and health, they know surprisingly little about the major factors that determine the level of social support perceived by people. Individuals' perceptions, that they have social support from one or more other persons is the outcome of an interpersonal transaction in which the emitted behaviors and acts of others are perceived by them and interpreted as supportive or nonsupportive. Except in the case of formal support or care givers, this transaction is part of their ongoing, mutual relationships in which they both give and receive support. Individuals' perception of support may change as a result of what others actually do toward them or as a result of how they perceive and interpret these acts of others.

For a number of reasons, this chapter has focused primarily on factors that may affect what others do toward an individual (or the individual toward them) rather than on factors influencing how these acts are perceived. I suspect that people who perceive a lack of support do so primarily because they lack supportive inputs rather than that they misperceive or misinterpret the inputs they are getting. The major cutting point in research on effects of support has been between people who report virtually no supportive relationship and those with one or more such relationships (see Chapter 3). Efforts to enhance social suport are likely to be directed to increasing the input of supportive acts toward others. Thus, it seems reasonable to focus initially on factors affecting supportive inputs. If future research and experimental interventions suggest that these factors have limited effects, more attention can and should be paid to the individual and interpersonal characteristics that may prevent people from effectively perceiving and using the supportive inputs available to them.

The major determinants of the socially supportive inputs provided to individuals by others, are (1) their ability and motivation to provide support and (2) the degree to which the larger interpersonal and social context condones and supports such efforts. It is clear that people can learn to provide social support to others. What is less certain is the extent to which work organizations and the broader society normally provide people with the necessary skills and facilitate the giving of support. The instrumental, often competitive, orientations of United States society and work organizations may mitigate against support. Both the values and the structure of work organizations affect the quality of supportive relations within them. Social support is a critical element in human relations theories of organization, but the acceptance and implementation of these practices, especially by supervisors, appears to depend on the attitude of higher management. The structure and technology of work (for example, isolation of workers from interaction with others) has been shown to adversely affect coworker support.

Nonwork support appears to vary as a function of (1) the general level of social and family integration in an area, (2) the characteristics of specific family and friendship networks, and (3) the degree to which the work roles of people facilitate or hinder the development of nonwork relationships. Although hard evidence is scarce, it appears that nonwork support may be more readily and immediately available in smaller communities than in large cities. The changing nature of marital and family relationships, especially the increasing social equality between the sexes and rising rate of two-worker households has and will have important effects on family support systems, although the nature of these effects is still to be elucidated. Paradoxically, probably the best documented statement researchers can make about the determinants of nonwork support is that it is adversely affected by work stress. In sum, this chapter has presented some evidence, and much speculation, about the determinants of social support. Research on this topic is sorely needed. Nevertheless, Chapter 6 will argue that although knowledge of the effects and especially determinants of social support is limited in many ways, we can and should attempt to enhance social support at work, and view such efforts as experiments that will both advance the understanding of social support while putting that understanding to practical use.

6

WHAT SHOULD AND CAN BE DONE?

Stress management has become a growth industry in its own right. Individuals and organizations are increasingly undertaking preventive programs of exercise, meditation, relaxation, or biofeedback in an effort to reduce stress or its deleterious impact on health (see McLean, 1979, Chapter 8). These programs are generally in addition to the growth of professional therapeutic resources for treatment of stress-related health problems including alcohol and drug abuse. Almost all of these preventive and therapeutic programs are focused on the individual. That is, the preventive and therapeutic regimens of these programs center on changing the individual's behavior and pattern of responding to stressful situations. These individually oriented programs can be of substantial value, but they also have a substantial limitation or blind spot: they fail to recognize and respond to the role of the social environment in producing stress and responses to stress.

Although stress is ultimately in the eye of the beholder, individual's perceptions of stress represent responses to their social environments. Thus, a balanced and effective approach to stress reduction and stress management must attempt to modify characteristics of social environments as well as characteristics of individuals. It should also attempt, where possible, to improve the compatibility or fit between individuals and their social environments. What is one person's meat is sometimes another's poison.

The availability of social support is a critical feature of any individual's social environment and a potentially effective lever for making social environments less stressful, more healthful, or more conducive to effective adaptation to stress. This chapter will argue that steps can and should be taken to enhance social support so as to reduce work stress, improve health, and buffer the impact of stress on health. Chapters 2 to 5 provide the foundation of scientific knowledge upon which such efforts are based. In many ways this knowledge base is limited. More research is clearly needed. However, weighting the substantial potential benefits against the small costs and negligible risks involved, I think efforts to improve social support are equally or more justified as a response to the very real problem of occupational stress than programs of exercise, meditation, relaxation, or biofeedback. Ideally, research and practical application should take place in conjunction with each other, making both more useful and more effective.

WHAT SHOULD BE DONE?

Stress reduction, stress management, and social support

Social support is, however, only one element, although a promising and potentially important one, in a larger strategy aimed at reducing work stress, improving worker health, and generally enhancing the quality of work and life. While social support, at least from work-related sources, appears to reduce the experience of occupational stress, there are many other approaches to stress reduction, some of which are more direct and effective than enhancing social support. Support, however, has the additional capacity to buffer the impact of stress on health. Thus, for some, enhancing support may appear to be an alternative to other strategies of direct stress reduction. They would argue that work organizations need not try to reduce stress levels if they can protect people against the deleterious effects of stress by providing social support. To accept this line of argument would be a serious mistake. Yet I fear that some of the current enthusiasm for social support stems from a desire to avoid the often more difficult problems of restructuring work organizations and environments to reduce levels of occupational stress.

It is usually politically easier to cope with problems by adding new and desirable elements to a system than to modify previously established structures and processes. Thus, politicians prefer to meet the energy crisis with measures that increase supplies of energy rather than with measures to reduce demand by forcing greater conservation of energy. And problems of poverty and resource distribution are confronted more easily in times of prosperity when the disadvantaged can be helped by receiving a greater share of new income and wealth rather than by redistributing existing income and wealth. Similarly, it is easier to deal with the problem of work stress and its adverse effects on health by enhancing social support than by reducing stress.

There is something to be said for pursuing politically more expedient policies rather than less expedient ones. But we need to recognize that a great deal of occupational stress and the deleterious consequences thereof can and should be addressed by direct efforts at stress reduction, such as modifying stress-inducing organizational roles, practices, and policies to reduce or eliminate their stressful features. Lennart Levi's (1980) volume in this series discusses how changes can and should be instituted in societies and organizations to reduce levels of social and occupational stress. Efforts to enhance social support are *no* substitutes for such efforts.

The social support strategy should be seen as a means for reducing those stressors that are most affected by support or buffering people against the deleterious health consequences of stresses that, for compelling ethical, economic, practical, or other reasons cannot be reduced. Social support is not now, nor will it ever be, a panacea for all problems of occuptional stress and health. But it can be an important and effective component of a *comprehensive* effort to reduce work stress and improve occupational health, both physical and mental.

Work versus nonwork support

Assuming we have done all that we can reasonably or feasibly do to alleviate occupational stress by other means, where should we direct efforts to enhance social support as a means of further alleviating stress and especially buffering the impact of stress on health? A fundamental choice is between work-related and nonwork-related sources of support. For a variety of reasons I believe work-related sources of support should be the major targets of efforts to enhance support. Scientifically, the evidence in Chapter

4 indicates that work supervisors and coworkers play a greater role than nonwork sources of support in both reducing occupational stress and alleviating its effects on health. Pragmatically, work-related sources of support appear easier to reach and influence. Work organizations provide ready-made channels of communication and influence for enhancing the supportiveness of significant others, and allow a large number of people to be affected by any support enhancement program. In contrast, nonwork sources of support are more diffused, less accessible, and probably less easily influenced. Thus, efforts to enhance work-related sources of support should prove more cost efficient than efforts to enhance nonwork sources of support.

Finally, there are ethical reasons to focus on work-related sources of support in trying to alleviate occupational stress and its effects on health. To expect that spouses or friends and relatives will buffer workers against the deleterious effects of occupational stress is to displace onto the spouse and/or family and friends responsibilities that are rightfully those of the organization or the individual worker. In the United States at least, family, friendship, and especially marital relationships have too often been assumed to operate in the service of individual or organizational work achievement, with little attention paid to the deleterious effects that work organizations and involvements may have on the family. Since work stress appears to adversely affect nonwork sources of support (see Chapter 5), reducing work stress and enhancing social support within work organizations should have beneficial effects on familial, marital, and friendship relationships outside of work.

Nevertheless, research clearly suggests that friends and relatives outside the work place, most notably spouses, can be important sources of support in the face of occupational stress. Hence they cannot be ignored from an applied perspective. Certain types of potential occupational stress (shiftwork, mandatory overtime, intensive cyclical workloads, and deadlines) may impinge directly on family functioning as well as on the individual worker (Mott et al., 1965; Kanter, 1977b). In such cases spouses and other family members may need support themselves as well as constituting sources of support for the workers; and groups of similarly situated or affected spouses, couples, or families might be formed both to support each other and to share and develop modes of providing support for affected workers. With respect to the more usual types of occupational stresses that affect the worker primarily and the family only indirectly, there is no clear and obvious route to enhancing the mutual supportiveness of spouses toward each other. Workers and spouses might be given options to participate in programs aimed specifically at enhancing their capacities for helping each other deal with work-related stresses, or such concerns might be more directly incorporated into other educational programs directed at improving marital and family relationships. I expect the need for such programs has grown and will continue to grow as both members of marital couples are increasingly likely to be working outside the home.

The major thrust of efforts to manage stress at the level of organizational environment, however, must be directed toward: (1) reducing wherever possible, levels of work stress that deleteriously affect health, and (2) enhancing social support from work-related sources to help alleviate those stressors and stress reactions that cannot otherwise be reduced. The remainder of this chapter explores the likely benefits that will accrue from efforts to enhance social support at work and explores practical directions and methods for actually increasing social support.

Why enhance support at work?

Hard-headed, pragmatic readers may well be wondering whether enhancing levels of social support will actually reduce work stress, improve health, or buffer the impact of stress on health, and whether such efforts to enhance support detract from more central organizational goals such as productivity. To my knowledge, no one has systematically tried to both enhance social support and measure the success of their efforts in terms of actual perceptions of social support or levels of stress and health. Thus, no guarantees exist that efforts to enhance social support will always lead to the desired effects on stress and health. However, all of the evidence in Chapters 3 and 4 suggests that increasing levels of social support at work should have such effects.

Furthermore, I would argue that efforts to enhance social support are likely to have a variety of beneficial effects on organizations and individuals besides the effects on work stress and health that are the central concern of this book. A large body of organizational theory and research focuses on the importance of social support as a major contributor to worker morale and satisfaction and organizational effectiveness, quite apart from the effects of support on stress and health. That is, social support is likely to make workers happier and more productive, regardless of its effects on stress and health.

Major organizational theorists give high priority to social support at work. As we noted in Chapter 5, Rensis Likert (1967), founder of the Institute for Social Research and a noted organizational theorist and consultant, posits the "supportiveness and openness of leaders" as the first characteristic of his ideal type of organization (called System 4). Cohesive and supportive work groups also are critical elements in his theory of what makes organizations effective. The Scandinavian experiments in improving the quality of work through the use of semiautonomous work groups were designed to fulfill six basic needs of individuals including "social support and recognition." Empirical tests of Likert's theory and the results of the Scandinavian experiments provide considerable evidence that social support beneficially affects both workers' morale and organizational productivity, aside from any effects it may have on work stress or health.

In reviewing the literature on leadership in organizations, Katz and Kahn (1978, pp. 559–569) conclude that the two major functions of organizational leadership vis à vis subordinates are "task direction" and "psychological supportiveness." Research in the laboratory and in organizations shows that, in most situations, a combination of both of these functions is necessary to achieve high levels of productivity and of other indicators of organizational effectiveness. Again, social support contributes to the achievement of important individual and organizational goals other than reducing stress or improving health. If, as suggested in Chapter 5, Japan provides an example of an especially supportive society both inside and outside of work, it clearly indicates that social support at work is not inconsistent with organizational productivity and effectiveness, and may in fact be a major stimulus to such productivity and effectiveness.

A supportive work environment is also likely to enhance the effectiveness of other stress-management programs or therapeutic regimes directed at chronic, stress-related illnesses. A number of studies document the impact of social support in promoting adherence to therapeutic regimes. Social support enhances, for example, the likelihood of individuals maintaining their use of blood pressure medication (Caplan et al., 1976). Part of the success of exercise programs may depend on the extent to which they allow for, or facilitate, the giving and receiving of social support. Support is also important in efforts to reduce smoking, alcohol and drug abuse, and other stress-related problems. Again, enhancing social support is likely to have beneficial effects for workers and organizations beyond any direct effects in reducing stress or promoting health.

Finally, the costs of efforts to enhance social support are relatively low, and they entail no major risks. To my knowledge enhancing levels of social support has never hurt anyone. Nor does it require elaborate equipment, take substantial time away from work activity, or require elaborate training beyond that which organizations already offer, or might offer, for other reasons. Since many theories of organizational effectiveness emphasize the importance of supportive leadership and supportive group interaction, much work in organizational development involves efforts to enhance social support. Organizations already involved in such programs could use them as part of a larger effort at enhancing social support in the work setting. Organizations that have not experimented with programs for increasing social support, ought to do so for the good of the organizations as well as the good of their employees.

Although many of my scientific colleagues would rightfully emphasize the limitations in existing knowledge of social support, I find the evidence and arguments in favor of efforts to enhance social support quite compelling and the costs and risks of attempting such efforts quite minimal. Thus, I would hope that managers of work organizations will be willing to undertake such efforts, and that scientists will be willing to advise in and evaluate such efforts. The remainder of this chapter suggests what to do in this direction.

WHAT CAN BE DONE?

Building social support into organizations

Enhancing levels of social support at work involves more than giving workers access to professional or lay counselors whose job it is to provide psychological therapy and support. Special programs and personnel for dealing with problems of occupational mental health, alcoholism, and drug abuse are, of course, desirable and valuable in most work settings, sometimes essential. When I speak of enhancing levels of social support at work, however, I have in mind something more—making the giving and receiving of social support a central and normal part of the ongoing structure and process of work organizations. Health professionals and social scientists can play important roles in planning, education, and consultation activities toward this end, and they remain a critical resource for people with serious problems. If social support is to be effective in reducing stress, preventing health problems, and increasing workers' ability to adapt to the irreducible stresses at work, all people must be able to obtain support from the persons with whom they routinely work—superiors, subordinates, and coworkers or colleagues. These persons are at once more accessible, more familiar and similar in their experience and orientations, and more attuned to the unique problems of their work situation than any health professional or counselor can be. Thus, they must be the focus of efforts to enhance social support within work organizations.

The idea of making all employees more supportive toward each other, in an organization like General Motors with hundreds of thousands of employees, appears impossible and utopian. Yet the very organizational structures and processes that enable GM to coordinate all those people to produce cars can be harnessed to produce social support as well. This process can occur without great expenditure of time, resources, or energy, provided there is a strong commitment *at all levels of the organization* to the goal of institutionalizing social support into the normal structure and process of the organization.

As indicated in Chapters 3 and 4, social support from *one* significant other person can be quite effective in reducing occupational stress and especially in alleviating much, if not all, of the deleterious effects of occupational stress on health. Indeed, support from additional sources may have little or no additional benefit in many cases. Consequently, applied efforts at enhancing social support need only try to insure that as many workers as possible have a supportive relationship with one such significant other at work. Since one person can support a number of others, a company's energies and resources need not be diffused trying to make everyone supportive toward everyone else or to make sure every worker is enmeshed in a large network of significant relationships. The effort should be broad in covering as many workers as possible, but narrow in trying to enhance the supportiveness of one, or at most a few, key relationships involving each worker.

The key targets of efforts to build social support into the normal structure and process of work organizations must be persons who by virtue of their organizational positions have direct, face-to-face relations with a large number of other workers. The most obvious candidates are work supervisors or managers. Most workers have a supervisor or manager who is a significant person in their working lives. Most supervisors and managers have a number of subordinates (and a superior). Thus, enhancing the ability of supervisors or managers to provide social support, especially toward subordinates but also toward colleagues and superiors, allows an organization to enhance the amount of social support available to many workers by interventions involving a much smaller number. In unionized workplaces, each union steward similarly relates to a large number of union members. In those settings, then, union stewards are an alternative or complementary target of efforts to enhance levels of social support at work.

Because supervisors, managers, and stewards are involved in established channels of organizational communication and authority, they are accessible to influence through those channels. Knowledge of the importance of social support and of the skills necessary to provide support could and should be part of the basis of selecting, and/or procedures for training, people to be union stewards or supervisors and managers at all levels of the organization. This knowledge is especially necessary for first-line supervisors who often have more subordinates with whom they deal directly than supervisors at other levels. Some efforts are made to select and train supervisors and stewards so as to enhance their performance, and some of these efforts may already be consistent with efforts to enhance social support (for example, selection and training programs based on organizational theories that emphasize the importance of supportive supervision to both worker morale and organizational effectiveness). But, as Levi (1978) argues, organizations do much less of this kind of training than they could or should. Thus, organizations can and should initiate, or strengthen efforts to include social support in the process of supervisory (or stewards) selection and training.

Work supervisors cannot be the most significant source of support for all workers, however. Some workers do not have a supervisor or have one in only the most formal sense (for example, the self-employed or those in largely collegial firms or organizations). Further, some supervisors may have far too many subordinates to provide effective social support to all of them. Thus, in many work situations efforts to enhance social support must extend beyond supervisors. Even where supervisors are readily available, they may not necessarily be the most effective sources of support with respect to all types of work stress or health problems, and the status differential between supervisor and subordinate can be a significant impediment to forming a truly supportive relationship. Fostering supportive relationships among coworkers, then, must be another major target of efforts to enhance social support. Unions and occupational or professional associations, as well as employment organizations, can play important roles in this area.

Although efforts to enhance social support can be targeted on certain types of organizational roles, these efforts must occur in the context of a broader organization-wide effort and commitment. Workers will not support each other to the extent they could if the power and reward structures of the organization do not encourage (much less discourage) such efforts. Management can ask first-line supervisors to provide support to their subordinates, and give them training in doing so, but if the first-line supervisors receive little or no support from their superiors, or if they are evaluated and rewarded solely on the basis of the instrumental activity or productivity of their subordinates (especially viewed in the short run), they will have little motivation to support their subordinates. Similarly, if workers are not supported by their first-line supervisors, are constantly placed in competition with their peers, and are rewarded only for productive activity for which they are clearly personally responsible, they will have little motivation to provide coworker support. Thus, higher levels of an organization should constitute models and positive sanctions for the efforts of lower levels to be more supportive.

In sum, the fields of mental health and organizational development have the technologies for teaching people how to give and receive social support. I will explore aspects of that technology shortly. Whether researchers begin to experiment with and utilize that technology in organizational contexts, and whether the effects of such experiments will be positive, consequential, and enduring depends on the willingness of organizations, especially their leaders, to recognize enhanced social support, improved occupational health, reduced occupational stress, and improved quality of life as major goals alongside traditional concerns with organizational productivity, growth, and survival. These newer goals are not necessarily incompatible with, and may even promote the traditional goals. But the newer goals will be taken seriously and achieved only if the opportunity and reward structure of the organization is made consistent with their realization.

Making individuals more supportive toward each other

Suppose individuals wish to enhance their ability to give or receive social support, or an organization is prepared to commit itself to making its members more supportive of one another. What exactly could they or should they do? I am a researcher rather than a counselor or organizational change agent. Thus, I do not propose to offer a how-to manual for giving and receiving social support. Rather I would like to develop a number of basic principles, grounded in existing research that any effort to enhance social support must incorporate. The details of implementation are certainly within the competence of many kinds of human service workers.

Accessibility. The first principle seems obvious, but it is crucial and sometimes overlooked. That is, if individuals are to have a supportive relationship with another person, that person must be accessible to them both physically and psychologically. They must be able to communicate with that person relatively easily and frequently about issues and problems of concern to them. All existing research (Chapters 3 and 4) suggests that physical or social isolation, at work or elsewhere, is deleterious for health and markedly reduces an ability to adapt to stress. Thus, anyone concerned with improving the levels of social support experienced by self or others, must seek to insure that conditions of work facilitate, rather than impede, opportunities for relatively free and open communication between workers. For many workers, supervisors and union stewards can be effective sources of support, but only if they are accessible to the workers, and similarly for coworkers.

Factors that impede accessibility are many. The most obvious is physical isolation. Chapter 5 reviewed evidence that factory workers in highly individuated and demanding jobs experienced markedly lower levels of coworker support. But such conditions of physical isolation are not unique to factories. They also characterize many craft and service workers (repairers who work alone, lone police officers in patrol cars), sales and clerical workers (traveling salespeople, isolated clerks), and even high level professionals such as many physicians and dentists in solo private practice.

Combined with physical isolation can be social isolation—lack of someone with whom to effectively communicate about work stresses even if not physically isolated. Police officers, salesmen, and dentists see many clients or patients, but these contacts are more likely to be sources of stress than support. Just as changes in the structure and technology of work (for example, redesign of assembly-line production) can increase social support in factories, organizing police officers or salesmen into working pairs or teams, and dentists and physicians into group practices, could markedly enhance levels of social support, thus reducing some stresses and facilitating adaptation to others. Since these are all notoriously high-stress occupations with high rates of suicide, alcoholism, depression, coronary disease, and other stress-related disorders, the potential gains of enhancing social support are considerable. The tendency to fractionate and isolate jobs in the name of efficiency is great, but whatever short-run gains in productivity may occur are likely to be more than outweighed by the personal, organizational costs of increased health problems and performance lapses over time. Thus, whenever changes in work arrangements are considered, the potential effects on the accessibility of social support should be carefully evaluated. Where workers must of necessity be relatively isolated in doing their jobs, special efforts must be made to insure that supervisors or other potential supporters are as accessible as possible.

One other form of social isolation should be mentioned, which is increasingly common in work organizations today as the composition of the work force changes to include more women and minorities in all occupations. Being the only woman or minority person in a given occupational level or organizational unit can often be profoundly socially isolating. Such "tokens," in the words of Rosabeth Kanter (1977a), are both exposed to unique work stresses and largely cut off from others who are likely to understand and empathize with their situation. The development of women's and minority "networks" within organizations or professional associations represent a constructive response to these problems. Efforts must also be made to facilitate communication and support between majority and minority.

Accessibility, however, is only a necessary, but not a sufficient condition for the establishment of supportive relationships. Once people are accessible, they must behave toward each other in ways that produce feelings of social support. This behavior entails providing key personnel with the necessary orientation and skills to enhance social support.

Training. As indicated in Chapter 5, merely asking or telling work supervisors or others to be more helpful or supportive toward people with whom they work is not likely to be productive, and may even be counterproductive. Most people need some instructions or training to become more supportive. However, counseling programs in which lay persons received brief training and were then as effective as professional counselors in providing support suggest that the training task is a relatively modest one (Durlak, 1979).

Since social support has at least four aspects (see Chapter 2) —instrumental aid, information, appraisal, and emotional sustenance—training people to give social support could involve training in giving all of these. Emotional support, however, seems at once more consequential for stress and health, more general in its effects, and most difficult to transmit as a skill. Thus, it should be a focal point in efforts to improve social support.

Existing evidence reviewed in Chapters 3 and 4 clearly suggests that emotional support is the most clearly and generally beneficial form of social support with respect to stress reduction and health promotion. The effects of instrumental aid, appraisal, and information are likely to depend greatly on the exact nature of the aid, appraisal, or information and the context in which they are provided. In contrast, emotional support seems to be a general currency that has beneficial, and at worst benign, effects on a wide range of individuals in a wide range of settings. It may also condition the effects of the other forms of support. Honest appraisal, for example, is much easier to convey and accept in the context of an emotionally supportive relationship. Finally, where it is relatively easy for individuals to give aid, information, and perhaps appraisal, their providing emotional support is a somewhat more subtle process.

As indicated in Chapter 5, providing emotional support is not equivalent to cheerleading. The needed involvement is clearly indicated in Gottlieb's (1978) inventory of "emotionally sustaining behaviors," presented in Table 2.1 of Chapter 2. These behaviors comprise a relationship characterized by a person's empathic listening to the problems of another coupled with expressions of positive regard toward that other and a feeling of intimacy and trust on the part of the other. These characteristics are the very qualities that psychotherapist Carl Rogers (1961) uses to describe a "helping relationship," and has shown to be effective in psychotherapeutic situations and more generally. Further, as noted in Chapter 5, it is clear that the skills necessary to establish such relationships can be transmitted to lay people, such that these lay persons can function as effectively as professionals in providing certain types of counseling or support (Rioch et al., 1963; Durlak, 1979).

I emphasized at the beginning of this chapter that supportive behavior generally, and from supervisors in particular, is likely to not only ameliorate problems of work stress and health, but also enhance organizational effectiveness. This notion is confirmed in an extensive review of studies of superior–subordinate communication by Jablin (1979). Jablin cites Redding's (1972) review of a series of studies of the characteristics of "good" versus "poor" supervisors as determined by higher management evaluation. The characteristics that Redding (1972, pp. 436–446, as summarized by Jablin, 1979, p. 1209) finds distinctive about good supervisors are very much like Gottlieb's "emotionally-sustaining behaviors" and Roger's helping relationship, though more focused on job activities per se:

1. *The better supervisors tend to be more "communication-minded"; for example, they enjoy talking and speaking up in meetings; they are able to explain instructions and policies; they enjoy conversing with subordinates; . . .*
2. *The better supervisors tend to be willing, empathic listeners; they respond understandingly to so-called silly questions from employees; they are approachable; they will*

listen to suggestions and complaints with an attitude of fair consideration and willingness to take appropriate action; . . .

3. *The better supervisors tend (with some notable exceptions) to "ask" or "persuade," in preference to "telling" or "demanding"; . . .*
4. *The better supervisors tend to be sensitive to the feelings and ego-defense needs of their subordinates; for example, they are careful to reprimand in private rather than in public; . . .*
5. *The better supervisors tend to be more open in their passing along of information; they are in favor of giving advance notice of impending changes, and of explaining the "reasons why" behind policies and regulations.*

(REDDING, 1972)

The dissemination and acquisition of these skills of effective listening, responding, communicating, and caring must be the core of individual or organizational efforts to become more socially supportive.

Rewards and reinforcement. If such training is to have a sustained positive impact on supervisors or other organizational members, however, the acquisition and continued *use* of these skills must be rewarded and reinforced at all levels of the organization. Such reinforcement should not be difficult since such behavior generally produces a variety of positive effects on the individuals involved and for the organization as a whole. Nevertheless, I am often struck that the lack of such skills and/or the failure to utilize them is one of the most common sources of organizational problems. The United States tends to be a very pragmatic, instrumental society, and especially among men who dominate most work environments, the kinds of skills (empathy, consideration, sensitivity) involved in giving or receiving social support, are often devalued as "soft," "nonproductive," or "nonmasculine." In most organizational settings such behaviors are not likely to be as naturally and easily rewarded and reinforced as people might expect and hope. I would, therefore, reiterate again the need for organization-wide commitment to such goals, especially at higher levels of management.

Strategic focus. Where resources are scarce, initial efforts to enhance work-related social support should be directed at those workers or job contexts characterized by high levels of work stress or health problems. Within this set of persons or situations, choose those situations most deficient in levels of social support. That is, social support should be applied where it is likely to do the most good in reducing work stress or buffering workers against the deleterious health consequences of stress that cannot be reduced. Such focal persons or situations are often identifiable from general, or preferably organizationally specific, knowledge of occupational or job-context differences in stress or health. Workers undergoing predictable work-related transitions or crises (job entry, job change, job loss, retirement), most of which are at least potentially stressful, could also be priority targets for efforts to enhance work-related support. However, remember that work stress is largely in the eye of the beholder (McMichael, 1978), and try to reach as many workers as possible since some people will experience significant stress under conditions that are rather benign for many people.

Combining research and practice. Initial efforts to use social support as a mechanism for improving health and reducing stress must necessarily be somewhat experimental. Hence they should involve collaboration between scientists and organizational and health practitioners. Any program of planned intervention in an organization should include the following elements:

1 Initial assessment of levels of social support, work stress, health and well-being—these data can serve both as a source of information for planning and targeting intervention efforts and as a baseline for evaluating the effects of intervention programs.

2 Planning and implementation of a program of organizational change that is systemic in nature even if intended to primarily affect only a portion of the organization's members—the influence of technology, organizational structure, and educational effects should be as mutually compatible as possible. Involvement of the target population in the planning process is essential to the effectiveness of the program.

3 A careful evaluation of the impact of the intervention effort on supportive behavior and levels of perceived support, and of both of these on work stress, health, and the equality of work—

the results of this evaluation should contribute both to the scientific knowledge and to planning and implementation of further social support interventions.

The interest and attention that the concept of social support has recently received among both scientists and laypersons is itself indicative, I think, of the degree to which people perceive unmet needs for social support. One expression of this need has been increased concern with strengthening the family. Some have advocated, for example, that family impact statements be prepared with respect to new social policies. This idea might be broadened to "social support impact statements." Work-related social support is clearly a major area for research and action to understand and enhance social support, as a means to reducing work stresses and enhancing the quality of work and life. Indeed, work organizations can be both potent mechanisms for achieving planned social change and excellent laboratories for research and evaluation. However, this potential will require commitment and change at the organizational, as well as individual level.

CONCLUSION

I argued at the beginning of this chapter that support is not a panacea for the problems of occupational stress. Others at work (supervisors) can be sources of both stress and support. Even an individual's most supportive friends and associates can cause that individual considerable distress at other times. Further, all people vary in their needs for social support, and some people may even feel they already have enough or too much support. Consequently, efforts to enhance work-related social support are not always the most efficient and effective way of reducing occupational stress, promoting occupational health, or buffering the impact of stress on health.

In general, however, I think a strong case can be made that levels of social support experienced by most people in the United States are less than they could or should be. And I have tried to show that it is possible to make levels of support more adequate, at least at work, and that such efforts are likely to have a beneficial impact on individual and organizational well-being. Organizations must begin to move in this direction—carefully planning, assessing, and evaluating what they do. The outcome is potentially both a better science of work stress and support due to its grounding and application in everyday lives and a better quality of everyday work and life due to its grounding in a solid body of scientific theory and research.

People have relied for a long time on material technology as a means for improving the quality of life. More recently they have turned to psychological technologies (psychotherapy, personal growth, and development training) in an effort to provide the sense of well-being that material technology has failed to give them. Here too, the results at best are mixed. It is time to recognize the quality of social relations as a major, perhaps the major, determinant of well-being and to develop and use technologies for improving them. Improving the quality of social support in work organizations and settings can be an important step in this direction.

APPENDIX A

BUFFERING VS. MAIN EFFECTS OF SUPPORT: A REGRESSION APPROACH

The discussion of buffering versus main effects of social support in Chapters 1 to 4 can be stated more concisely, but also more technically, in terms of a set of linear equations, such as one would estimate via multiple regression. The basic model can be graphed as follows (essentially Fig. 2.1):

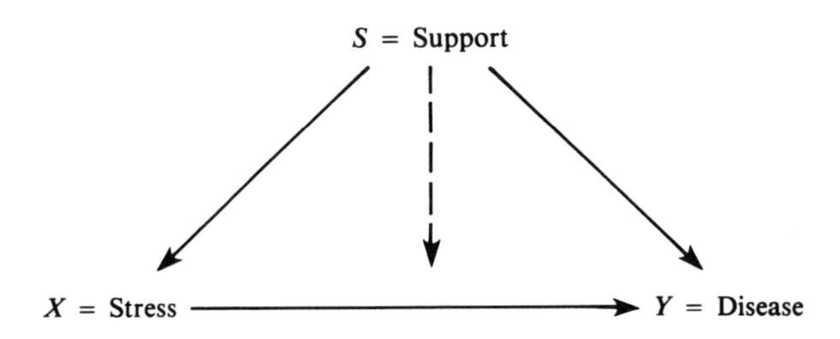

where a dashed line indicates an interactive or conditioning effect. Assuming all of the additive effects are linear, this model can be written in the form of two equations:

$$\hat{X} = a + bS \quad (1)$$

$$\hat{Y} = a + b_1X + b_2S + b_3SX \quad (2)$$

That is, stress (X) is a linear function of support and disease (Y)

is a linear function of stress (X), support (S), and their multiplicative product (SX).

Equation (1) merely says that support can have a main effect on stress. Equation (2) indicates that the effects of stress and support on health may be additive or main effects (if b_3 is not significantly different from zero) or interactive or buffering effects (if b_3 is significantly different from zero). Thus, the first crucial test for a buffering effect of support on the relationship between stress and health is whether the b_3 coefficient in a regression equation such as equation (2) is significantly different from zero. If it is, then the estimated effect of stress (X) on disease (Y) varies across levels of support as follows:

$$\text{if } S = 0 \quad \hat{Y} = a + b_1X \tag{2a}$$

$$\text{if } S = 1 \quad \hat{Y} = a + b_1X + b_2(1) + b_3(1)X = (a + b_2) + (b_1 + b_3)X \tag{2b}$$

$$\text{if } S = 2 \quad \hat{Y} = a + b_1X + b_2(2) + b_3(2)X = (a + 2b_2) + (b_1 + 2b_3)X \tag{2c}$$

•

•

•

$$\text{if } S = k \quad \hat{Y} = a + b_1X + b_2(k) + b_3(k)X = (a + kb_2) + (b_1 + kb_3)X \tag{2d}$$

These equations make clear the meaning of the coefficients in equation (2):

a = the estimated *intercept* of the regression of Y on X when $S = 0$;

b_1 = the estimated *slope* of the regression of Y on X when $S = 0$;

b_2 = the estimated *change* in the *intercept* of the regression of Y on X for each unit change in S; and

b_3 = the *change* in the *slope* of the regression of Y on X for each unit change in S.

Thus, b_1 of equation (1) estimates the main effect of support on stress, and *under certain assumptions and conditions* coefficient b_2 of equation (2) indicates whether support has a main effect on health or disease, and coefficient b_3 indicates whether support buffers the effect of stress on health. Specifically:

1 *If* all variables are scaled positively with a real zero point (that is, range from 0 up), b_2 of equation (2) estimates the main effect of support on disease. Theoretically, if b_2 is significantly different from zero, it should be negative since researchers expect that as support increases, rates of disease should decrease. A positive b_2 would indicate that support increases disease and hence disconfirms the general theory of social support and its effects.

2 Buffering is evidenced only if b_1 is significantly >0 (indicating that stress increases rates of disease when support is zero or low) and b_3 is significantly <0 (indicating that the effect of stress on health declines as support increases) and of a size such that the estimated slope of the regression of disease at the highest level of support is not significantly less than 0. That is, if support ranges from 0 to k, $b_1 + kb_3$ should be ≥ 0 (or nonsignificantly less than 0). Any other pattern of b_1 and b_3 would *not* be consistent with a buffering hypothesis.

It should be stressed that the only safe and sure way to make sure that a set of coefficients for equation (1) and (2) are consistent with the buffering hypothesis is to plot the expected regression lines on various levels of support by substituting appropriate values in equations such as (2*a*) to (2*d*).[1]

Note that using equation (2) to test for the main and buffering effects of support on disease or health assumes that both of these effects are linear: the change in health or in the slope of the regression of stress on health is the same for each unit change in support. This assumption seems reasonable, is consistent with current theoretical statements about the effects of support on stress and health, and is consistent with empirical results of analyses by House and Wells (1978) and LaRocco et al. (1980). However, it is both theoretically and empirically possible that such effects are not always linear, but have thresholds. For example, increases in support from low to moderate levels might have substantial main effects and buffering effects with respect to health, but increases in support beyond a moderate level may be of little benefit. Such effects would be more accurately estimated if the effects of support are allowed to be nonlinear.

Allow for all possible nonlinear effects of support by simply

---

1. The same is true with respect to equations (3) and (3*a*) to (3*d*).

coding support as a set of dummy variables ($S_l = 1$ if support is "low," 0 otherwise; $S_m = 1$ if support is "medium," 0 otherwise; $S_h = 1$ if support is "high," 0 otherwise). The analyses reported in Chapter 4 began this way, and moved to using multiplicative interaction terms after determining that there were no threshold effects. Omitting one dummy variable to avoid linear dependency, equation (2) would be rewritten as:

$$\hat{Y} = a + b_1X + b_2S_m + b_3S_h + b_4S_mX + b_5S_hX \qquad (3)$$

As with equation (2), values of the support variables can be substituted in equation (3) and the estimated regressions of disease on stress derived for each category of support.[2]

$$\text{if } S_l = 1 \text{ (hence } S_m = S_h = 0) \quad \hat{Y} = a + b_1X \qquad (3a)$$

$$\text{if } S_m = 1 \text{ (hence } S_l = S_h = 0) \quad \hat{Y} = a + b_1X + b_2(1) + b_4(1)X = (a + b_2) + (b_1 + b_4)X \qquad (3b)$$

$$\text{if } S_h = 1 \text{ (hence } S_l = S_m = 0) \quad \hat{Y} = a + b_1X + b_3(1) + b_5(1)X = (a + b_3) + (b_1 + b_5)X \qquad (3c)$$

These equations clarify the meaning of the coefficients in equation (3):

a = the *intercept* of the regression of Y on X under low support

b_1 = the *slope* of the regression of Y on X under low support

b_2 = the *difference* between the *intercept* for the regression of Y on X when support is medium versus low

b_3 = the *difference* between the *intercept* for the regression of Y on X when support is high versus low

b_4 = the *difference* between the *slope* of the regression of Y on X when support is medium versus low

b_5 = the *difference* between the *slope* of the regression of Y on X when support is high versus low.

---

2. These estimates are identical to those that would be obtained by estimating separate regressions of health on stress for people with low, medium, and high support respectively. In contrast the estimates derived from a multiplicative interaction (SX) term only approximate those that would be obtained from a series of separate regressions, since the b_3 coefficient in equation (2) represents the average estimated unit change in the slope of the Y on X regression across all units of the support measure.

Just as with the discussion of equation (2) only certain values of these coefficients would be consistent with current theoretical ideas that support improves health and/or buffers the impact of stress on health. Specifically, b_2 and b_3 should be significantly negative or nonsignificantly different from 0 (with b_3 not significantly less than b_2). Similarly, b_5 should be significantly negative (with $b_5 + b_1$ not significantly less than 0 and b_4 not significantly more negative than b_5).

APPENDIX B

PROBLEMS IN DETECTING CONDITIONING OR BUFFERING EFFECTS IN CROSS-SECTIONAL STUDIES

Chapters 3 and 4 briefly indicated that, compared to experimental and quasi-experimental designs, cross-sectional studies are not well suited for detecting the buffering effects of social support, or more generally what were termed conditioning effects in Fig. 1.2 of Chapter 1. Since this point is not generally appreciated, and was not appreciated by me until recently, let me briefly indicate why cross-sectional designs will find it easier to detect main or additive effects of social support than to detect buffering effects. Since the point is a general one, it will be stated here in terms of conditioning effects in general, the buffering effects of social support being one kind of conditioning effect.

Theoretically, evidence of conditioning effects derives from evidence of statistical interaction. However, conditioning variables may have main or additive effects on a dependent variable apart from any interactive effects they may have. Such main or additive effects of conditioning variables may be interpreted in several ways. First, they can be interpreted as simply additional causes of the dependent variable. For example, most symptoms of ill health decrease with social support, but the conditioning effects of social support will be over and above, and quite independent of, the main effects noted here.

Second, a main effect of a conditioning variable on a dependent variable may be the result of a conditioning effect, the independent variable of which has been omitted from the analysis. For example, if C conditions a relationship between X and Y, but X is not measured and/or not analyzed in conjunction with Y and C, then a main effect of C on Y may be observed rather than a conditioning effect of C on the $X \rightarrow Y$ relationship.

Finally, consider what will be manifest empirically when a conditioning variable (C) conditions a relationship ($X \rightarrow Y$) that is in the middle of a causal chain:

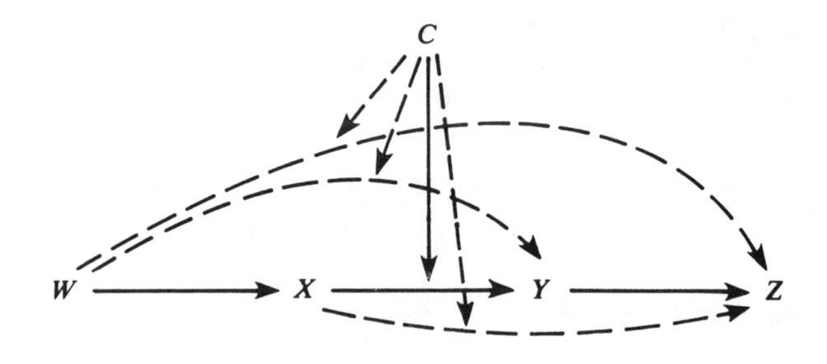

How would C affect other variables and relationships in the chain? First, C will have no necessary conditioning effect on other direct causal relationships in the chain, since these are either causally prior ($W \rightarrow X$) or subsequent ($Y \rightarrow Z$) to the effect of C. In contrast, C should condition (as indicated by the *dotted* arrows) all indirect causal relationships (indicated by the *dashed* arrows) in which the $X \rightarrow Y$ relationship is one link (that is, $W \rightarrow Y$, $X \rightarrow Z$, or $W \rightarrow Z$), although the conditioning effect may be attenuated (and even statistically nonsignificant) depending on how strong the links are in the causal chain and how C affects other relationships in the chain.

Importantly, however, the conditioning variable should be associated with all variables in the causal chain subsequent to the point of impact of the conditioning variable. In this example, C should be correlated with Y and Z (unless, of course, C also conditioned the $Y \rightarrow Z$ relationship in such a way as to counteract its earlier effects on Y). The sign, strength, and significance of this relationship will depend on the nature and strength of the conditioning effect of C on the $X \rightarrow Y$ relationship and the strength of the $Y \rightarrow Z$ relationship.

These considerations are especially important in a cross-sectional study. W, X, Y, and Z in the preceding diagram may represent not only conceptually distinct variables with a causal ordering (for example, job characteristics → perceived stress → general job evaluations → health outcomes); they may also represent measures of the same variable at two or more points in time. For example, a conditioning variable may act to modify an initial perception of stress in response to objective job characteristics, which will in turn alter health outcomes ($W =$ job characteristics → $X =$ perceived stress at time 1 → $Y =$ perceived stress at time 2 → Z health). If the cross-sectional study assesses perceived stress at time 1 rather than time 2, the conclusions drawn as to the presence or absence of statistical interactions involving C will be different. With a time 1 measure of perceived stress no conditioning effect of C on the relationship of job characteristics to perceived stress (i.e., $W \rightarrow X$) would be observed, but C would condition the impact of perceived stress on health ($X \rightarrow Z$). An exactly opposite pattern of effects occurs if perceived stress is measured at time 2.[1]

The moral of all this is that if significant interactions in cross-sectional studies are found between a putative conditioning variable (C) and an independent variable in predicting a dependent variable, researchers can be reasonably confident that C does exert true conditioning effects. But if researchers do not find such interaction effects, the inference task is not as easy. They must also pay attention to direct and indirect (that is, mediated by some other independent variable) main or additive effects of the conditioning variable on the dependent variable, as these *may* be evidence of true conditioning effects that are not adequately captured in the study due to the omission of an independent variable in a conditioned relationship or to the time of measurement of the indepen-

1. These problems are inherent to cross-sectional studies but not unique to them. Longitudinal designs do not necessarily resolve the dilemma since perceived stress might be measured, for example, at two points in time, but both are prior (or subsequent) to the impact of the conditioning variable. When the stresses at issue are chronic ones, as with many perceived work stresses, it may be difficult in some cases to observe buffering. It should be easier if the stress measure is a unique event (a job change) or an unchanging objective characteristic of the environment. On the other hand, in some cases buffering may be a relatively continuous process, as when social support allows persons to recognize that chronic job stresses are not their fault, thus lessening the impact on their self-esteem and health. Many of the buffering effects observed in the cross-sectional analyses in Chapter 4 are probably of this type.

dent or dependent variable in a conditioned realtionship. (They may also merely indicate an independent, direct effect of the conditioning variable on the dependent variable.) Of course, the absence of both additive (main) and interactive (conditioning) effects of a putative conditioning variable would constitute strong evidence against its having conditioning effects or direct effects on the dependent variable.

Because cross-sectional studies are weak designs for detecting conditioning or buffering effects, I find the buffering effects found in the cross-sectional studies reviewed in Chapter 4 particularly striking. Still, researchers will adequately understand the buffering effects of social support only when they study them in more appropriate longitudinal and field-experimental studies.

REFERENCES

Back, K. W., and M. D. Bogdonoff (1967). Buffer conditions in experimental stress. *Behavioral Science* **12**:384–390.

Bakke, E. W. (1940). *Citizens Without Work.* New Haven: Yale University Press.

Berg, I., M. Freedman, and M. Freeman (1978). *Managers and Work Reform: A Limited Engagement.* New York: Free Press.

Berkman, L. F. (1977). Social networks, host resistance and mortality: A follow-up study of Alameda County residents. (Unpublished doctoral dissertation.) Berkeley: University of California.

Berkman, L. F., and S. L. Syme (1979). Social networks, host resistance, and mortality: A nine-year follow-up study of Alameda County residents. *American Journal of Epidemiology* **109**(2):186–204.

Blauner, R. (1964). *Alienation and Freedom.* Chicago: University of Chicago Press.

Bovard, E. W. (1959). The effects of social stimuli on the response to stress. *The Psychological Review* **66**(5):267–277.

Bradburn, N. M. (1969). *The Structure of Psychological Well-Being.* Chicago: Aldine.

Brown, G. W., M. N. Bhrolehain, and T. Harris (1975). Social class and psychiatric disturbance among women in an urban population. *Sociology* **9**:225–254.

Caplan, G. (1974). *Support Systems and Community Mental Health.* New York: Behavioral Publications.

Caplan, G., and M. Killilea (1976). *Support Systems and Mutual Help.* New York: Grune & Stratton.

Caplan, R. D. (1979). Social support, person-environment fit, and coping. Pp. 89–138 in L. A. Ferman and J. P. Gordus (eds.), *Mental Health and the Economy*. Kalamazoo, Mich.; The W. E. Upjohn Institute for Employment Research.

Caplan, R. D., S. Cobb, J. R. P. French, R. V. Harrison, and S. R. Pinneau (1975). *Job Demands and Worker Health*. U. S. Dept. of Health, Education and Welfare, *HEW Publication No. (NIOSH) 75-160.*

Caplan, R. D., E. A. R. Robinson, J. R. P. French, Jr., J. R. Caldwell, and M. Shinn (1976). *Adhering to Medical Regimes: Pilot Experiments in Patient Education and Social Support*. Ann Arbor: Institute for Social Research.

Cassel, J. (1970). Physical illness in response to stress. Pp. 189–209 in S. Levine and N. Scotch (eds.), *Social Stress*. Chicago: Aldine.

Cassel, J. (1976). The contribution of the social environment host resistance. *American Journal of Epidemiology* **102**(2):107–123.

Chen, E., and S. Cobb (1960). Family structure in relation to health and disease. *Journal of Chronic Diseases* **12**(5):544–67.

Coates, D. and C. B. Wortman (1980). Depression maintenance and interpersonal control. Pp. 149–181 in A. Baum and J. Singer (eds.), *Advances in Environmental Psychology: Applications of Personal Control*. Vol. II. Hillsdale, N.J.: Lawrence Earl Baum Associates, Inc.

Cobb, S. (1976). Social support as a moderator of life stress. *Psychosomatic Medicine* **38**(5):300–314.

Cobb, S. (1979). Social support and health through the life course. Pp. 93–106 in M. W. Riley (ed.), *Aging from Birth to Death: Interdisciplinary Perspectives*. Washington, D.C.: American Association for the Advancement of Science.

Cobb, S., and S. V. Kasl (1977). *Termination: The Consequences of Job Loss*. U. S. Dept. of Health, Education and Welfare, *HEW Publication No. (NIOSH) 77-224.*

Conger, J. J., W. L. Sawrey, and E. S. Turrell (1957). The role of social experience in the production of gastric ulcers in hooded rats placed in a conflict situation. Based on paper presented at the Annual Meeting of the American Psychological Association, New York City.

Durlak, J. A. (1979). Comparative effectiveness of paraprofessional and professional helpers. *Psychological Bulletin* **86**:80–92.

Epley, S. W. (1974). Reduction of the behavioral effects of aversive stimulation by the presence of companions. *Psychological Bulletin* **81**(5):271–283.

Festinger, L. (1954). A theory of social comparison processes. *Human Relations* **7**:117–140.

Fischer, C. (1973). Review of *A Nation of Strangers,* by Vance Packard. *American Journal of Sociology* **79**(1):168–73.

Fischer, C. (1976). *The Urban Experience.* New York: Harcourt Brace Jovanovich.

French, J. R. P., Jr., W. Rodgers, and S. Cobb (1974). Adjustment as person–environment fit. Pp. 316–333 in G. V. Coelho, D. A. Hamburg, and J. E. Adams (eds.), *Coping and Adaptation.* New York: Basic Books.

French, J. R. P., Jr., J. Israel, and D. Aas (1960). An experiment on participation in a Norwegian factory. *Human Relations* **13**:3–19.

Friedan, Betty (1963). *The Feminine Mystique.* New York: Dell Books.

Gore, S. (1973). The Influence of Social Support in Ameliorating the Consequences of Job Loss. Unpublished doctoral dissertation. University of Pennsylvania.

Gore, S. (1978). The effect of social support in moderating the health consequences of unemployment. *Journal of Health and Social Behavior* **19**:157–165.

Gottlieb, B. H. (1978). The development and application of a classification scheme of informal helping behaviours. *Canadian Journal of Science/Rev. Canad. Sci. Comp.* **10**(2):105–115.

Gove, W. R. (1972). The relationship between sex roles, marital status and mental illness. *Social Forces* **51**:34–44.

Gove, W. R. (1973). Sex, marital status and mortality. *American Journal of Sociology* **79**:45–67.

Gurin, G., J. Veroff, and S. Feld (1960). *Americans View Their Mental Health.* New York: Basic Books.

Hawley, A. H. (1973). Review of *A Nation of Strangers,* by Vance Packard. *American Journal of Sociology* **79**(1):165–68.

Heller, K. (1979). The effects of social support: Prevention and treatment implications. Pp. 353–382 in A. P. Goldstein and F. H. Kanfer (eds.), *Maximizing Treatment Gains: Transfer Enhancement in Psychotherapy.* New York: Academic Press.

Henry, J. P., and J. C. Cassel (1969). Psychosocial factors in essential hypertension. *Journal of Epidemiology* **90**(3):171–200.

Holmes, T. H., and R. H. Rahe (1967). The social readjustment scale. *Journal of Psychosomatic Research* **11**:213–218.

House, J. S. (1974a). Occupational stress and coronary heart disease: A review and theoretical integration. *Journal of Health and Social Behavior* **15**:12–27.

House, J. S. (1974b). Occupational stress and physical health. Pp. 145–170 in J. O'Toole (ed.), *Work and the Quality of Life: Resource Papers for Work in America.* Cambridge, Mass.: The MIT Press.

House, J. S. (1980). Occupational stress and the mental and physical health of factory workers. Ann Arbor, Mich.: Institute for Social Research Report Series.

House, J. S., and M. F. Jackman (1979). Occupational stress and health. Pp. 135–158 in P. Ahmed and G. Coelho (eds.), *Toward a New Definition of Health.* New York: Plenum Publishing.

House, J. S., and J. A. Wells (1978). "Occupational stress, social support, and health. In A. McLean, G. Black and M. Colligan (eds.), *Reducing Occupational Stress: Proceedings of a Conference.* DHEW (NIOSH) Publication **78-140**:8–29.

House, J. S., A. J. McMichael, J. A. Wells, B. N. Kaplan, and L. R. Landerman (1979). Occupational stress and health among factory workers. *Journal of Health and Social Behavior* **20**:139–160.

Jablin, F. M. (1979). Superior-subordinate communication: The state of the art. *Psychological Bulletin* **86**(6):1201–22.

Jaco, E. G. (1970). Mental illness in response to stress. Pp. 210–217 in S. Levine and N. A. Scotch (eds.), *Social Stress.* Chicago: Aldine.

Jahoda, M., P. Lazarsfeld, and H. Zeisel (1971). *Marienthal, the Sociography of an Unemployed Community.* Chicago: Aldine, Atherton.

Jones, E. E., and H. B. Gerard (1967). *Foundations of Social Psychology.* New York: Wiley.

Kagan, A. R., and L. Levi (1974). Health and environment—psychosocial stimuli: A review. *Social Science & Medicine* **8**:225–241.

Kahn, R. L., and T. Antonucci (1980). Convoys over the Life Course: Attachment, Roles and Social Support. In P. B. Baltes and O. Brim (eds.), *Life-Span Development and Behavior* (Vol. 3). Boston: Lexington Press.

Kahn, R. L., and D. Katz (1960). Leadership practices in relation to productivity and morale. Pp. 554–570 in D. Cartwright and A. Zander (eds.), *Group Dynamics: Research and Theory* (2d ed.). Evanston, Ill.: Row, Peterson.

Kahn, R. L., D. M. Wolfe, R. P. Quinn, J. D. Snoek, and R. A. Rosenthal (1964). *Organizational Stress: Studies in Role Conflict and Ambiguity.* New York: Wiley.

Kanter, R. M. (1977a). *Men and Women of the Corporation.* New York: Basic Books.

Kanter, R. M. (1977b). *Work and Family in the United States: A Critical Review and Agenda for Research and Policy.* New York: Russell Sage Foundation.

Kaplan, B. H., J. C. Cassel, and S. Gore (1977). Social support and health. *Medical Care* **25**(S, Supplement):47–58.

Kasl, S. V. (1974). Work and mental health. Pp. 171–196 in James O'Toole (ed.), *Work and the Quality of Life.* Cambridge, Mass.: MIT Press.

Kasl, S. V. (1978). Epidemiological contributions to the study of work stress. Pp. 3–48 in C. L. Cooper and R. Payne (eds.)., *Stress at Work.* New York: Wiley.

Katz, D., and R. L. Kahn (1978). *The Social Psychology of Organizations* (2d ed.). New York: Wiley.

Komarovsky, M. (1940). *The Unemployed Man and His Family.* New York: Dryden Press.

LaRocco, J. M., and A. P. Jones (1978). Coworker and leader support as moderators of stress-strain relationships in work situations. *Journal of Applied Psychology* **63**:629–634.

LaRocco, J. M., J. S. House, and J. R. P. French, Jr. (1980). Social support, occupational stress, and health. *Journal of Health and Social Behavior* **21**(September):202–218.

Lazarus, R. S. (1966). *Psychological Stress and the Coping Process.* New York: McGraw-Hill.

Levi, L. (1978). Psychosocial stress at work: Problems and prevention. Pp. 216–222 in A. McLean, G. Black, and M. Colligan (eds.), *Reducing Occupational Stress: Proceedings of a Conference.* DHEW (NIOSH) Publication **78-140**:8–29.

Levi, L. (1980). *Occupational Stress: Sources, Management, and Prevention.* Reading, Mass.: Addison-Wesley.

Levine, S., and N. Scotch (1970). *Social Stress.* Chicago: Aldine.

Liem, G. R. and J. H. Liem (1979). Social support and stress: some general issues and their application to the problem of unemployment. Pp. 347–379 in L. A. Ferman and J. P. Gordus (eds.), *Mental Health and the Economy.* Kalamazoo, Mich.: The W.E. Upjohn Institute for Employment Research.

Liddell, H. (1950). Some specific factors that modify tolerance for environmental stress. Chapter 7 in H. G. Wolff, S. G. Wolff, Jr., and

C. C. Hare (eds.), *Life Stress and Bodily Disease.* Baltimore: Williams and Wilkins.

Likert, R. (1961). *New Patterns of Management.* New York: McGraw-Hill.

Likert, R. (1967). *The Human Organization: Its Management and Value.* New York: McGraw-Hill.

Lin, N., R. L. Simeone, W. M. Ensel, and W. Kuo (1979). Social support, stressful life events and illness: A model and an empirical test. *Journal of Health and Social Behavior* **20**:108–119.

Lipowski, Z. J., D. R. Lipsitt, and P. C. Whybrow (1977). *Psychosomatic Medicine.* New York: Oxford University Press.

Lipset, S. M., M. Trow, and J. Coleman (1956). *Union Democracy.* New York: Free Press.

Lowenthal, M. F., and C. Haven (1968). Interaction and adaptation: Intimacy as a critical variable. *American Sociological Review* **33**(1): 20–30.

McGrath, J. E. (1970). *Social and Psychological Factors in Stress.* New York: Holt, Rinehart and Winston.

McLean, A. (1979). *Work Stress.* Reading, Mass.: Addison-Wesley.

McMichael, A. J. (1978). Personality, behavioural and situational modifiers of work stressors. Pp. 127–147 in R. Payne and C. Cooper (eds.), *Stress at Work.* London: Wiley.

Marmot, M. G., S. L. Syme, A. Kagan, H. Kato, J. B. Cohen, and J. Belsky (1975). Epidemiologic studies of coronary heart disease and stroke in Japanese men living in Japan, Hawaii and California: Prevalence of Coronary and Hypertensive Heart Disease and Associated Risk Factors. *American Journal of Epidemiology* **102**(6): 514–525.

Mason, J. W. (1975). A historical view of the stress field: Part I and II. *Journal of Human Stress* **1**(March):6–12 and (June):22–36.

Matsumoto, Y. S. (1970). Social stress and coronary heart disease in Japan: A hypothesis. *Milbank Memorial Fund Quarterly* **48**:9–31.

Mayo, E. (1933). *The Human Problems of an Industrial Civilization.* New York: MacMillan.

Mechanic, D. (1962). *Students Under Stress.* New York: Free Press.

Mechanic, D. (1970). Some problems in developing a social psychology of adaptation. Pp. 104–123 in J. McGrath (ed.), *Social and Psychological Factors in Stress.* New York: Holt, Rinehart and Winston.

Mitchell, J. C. (ed.) (1969). *Social Networks and Urban Situations.* Manchester: Manchester University Press.

Mott, P. E., F. C. Mann, Q. McLaughlin, and D. P. Warwick (1965). *Shift Work.* Ann Arbor: University of Michigan Press.

Nerem, R., M. J. Levesque, and J. F. Cornhill. Social environment as a factor in diet-induced atherosclerosis. *Science* **208**:1475–76.

Nuckolls, K. B., J. Cassel, and B. H. Kaplan (1972). Psychosocial assets, life crisis and the prognosis of pregnancy. *American Journal of Epidemiology* **95**:431–441.

Packard, V. (1972). *A Nation of Strangers.* New York: David McKay.

Pearlin, L., and J. Johnson (1977). Marital status, life strains, and depression. *American Sociological Review* **42**(October):704–715.

Pearlin, L., and C. Schooler (1978). The structure in coping. *Journal of Health and Social Behavior* **20**:200–205.

Pinneau, S. R., Jr. (1975). Effects of Social Support on Psychological and Physiological Stress. (Unpublished doctoral dissertation.) Ann Arbor: University of Michigan.

Pinneau, S. R., Jr. (1976). Effects of social support on occupational stresses and strains. Paper presented at the meeting of the American Psychological Association, Washington, D.C.

President's Commission on Mental Health (1978). *Report to the President.* Vols. I–IV. Washington, D.C.: U. S. Government Printing Office.

Raphael, B. (1977). Preventive intervention with the recently bereaved. Archives of *General Psychiatry* **34**:1450–1454.

Redding, W. C. (1972). *Communication Within the Organization: An Interpretive Review of Theory and Research.* New York: Industrial Communication Council.

Rioch, M. J., E. Elkes, A. A. Flint, B. S. Udansky, R. G. Newman, and E. Sibler (1963). NIMH pilot study in training mental health counselors. *American Journal of Orthopsychiatry* **33**:678–689.

Rogers, C. (1961). The characteristics of a helping relationship. Chapter 3 of C. R. Rogers, *On Becoming a Person.* Boston: Houghton, Mifflin.

Sarnoff, I., and P. G. Zimbardo (1961). Anxiety, fear, and social affiliation. *Journal of Abnormal and Social Psychology* **62**:356–363.

Schachter, S. (1959). *The Psychology of Affiliation.* Stanford, Calif.: Stanford University Press.

Seashore, S. E. (1954). *Group Cohesiveness in the Industrial Work Group.* Ann Arbor: Survey Research Center, Institute for Social Research, University of Michigan.

Selye, H. (1975). *The Stress of Life* (2d ed.). New York: McGraw-Hill.

Selye, H. (1976). Forty years of stress research: Principal remaining problems and misconceptions. *CMA Journal* **115**:53–56.

Sherif, M., and C. W. Sherif (1953). *Groups in Harmony and Tension.* New York: Harper & Row.

Slote, A. (1969). *Termination: The Closing at Baker Plant.* Indianapolis: Bobbs-Merrill.

Susser, M. (1967). Causes of peptic ulcer. *Journal of Chronic Disease* **20**:435–456.

Syme, S. L. (1974). Behavioral factors associated with the etiology of physical disease: A social epidemiological approach. *American Journal of Public Health* **64**:1043–1045.

Tajfel, H. (1969). Social and cultural factors in perception. Pp. 315–394 in G. Lindzey and E. Aronson (eds.), *Handbook of Social Psychology* (2d ed.), Vol. III. Reading, Mass.: Addison-Wesley.

Totman, R. (1979). *Social Causes of Illness.* New York: 1979.

Veroff, J., E. Douvan, and R. A. Kulka (1981). Forthcoming *The American Experience: A Self-Portrait over Two Decades.* Vols. I and II. New York: Basic Books.

Warshaw, L. (1979). *Stress Management.* Reading, Mass.: Addison-Wesley.

Webster's New World Dictionary (1959). *Webster's New World Dictionary of the American Language.* Cleveland: World Publishing Co.

Wells, J. A. (1978). *Social Support a Buffer of Stressful Job Conditions.* (Unpublished doctoral dissertation). Durham, N.C.: Duke University.

Whyte, W. H., Jr. (1956). *The Organization Man.* New York: Simon & Schuster.

NAME INDEX

SUBJECT INDEX